**Frontispiece** Adults of the mushroom sciarid (*Lycoriella auripila*, left) and the mushroom phorid (*Megaselia halterata*, right) showing essential differences in the two main pest species. The mushrooms show symptoms of *Verticillium fungicola* var. *fungicola* infection. (Flies not to scale)

# Mushrooms:
# Pest and Disease Control

# Mushrooms:
## Pest and Disease Control

J. T. Fletcher, P. F. White and R. H. Gaze

SECOND EDITION

Intercept
Andover, Hants

First published July 1986
Second edition published October 1989

**British Library Cataloguing in Publication Data**

Fletcher, J. T.
    Mushrooms : pest and disease control.
    I. Mushrooms, Edible — Diseases and pests
    I. Title    II. White, P. F.    III. Gaze, R. H.
    635'.8    SB608 .M9

    ISBN 0–946707–27–8

Typeset by Roger Booth Associates, Newcastle upon Tyne
Printed and bound in Great Britain by Athenæum Press, Newcastle upon Tyne

# Contents

# Colour Plates

1.  Wet bubble disease *(Mycogone perniciosa)* with typical amber droplets on the distorted mushrooms

2.  Dry bubble disease (*Verticillium fungicola* var. *fungicola*) distortion and cap spotting

3.  Dry bubble disease (*Verticillium, fungicola* var. *fungicola*) severe distortion of mushroom

4.  Cap spotting caused by (*Verticillium fungicola* var. *aleophilum*)

5.  Decay and discoloration of mushrooms resulting from Cobweb disease *Cladabotryum dendroides*

6.  Cap spotting caused by *Cladobotryum dendroides (Dactylium dendroides)*

7.  Cap spotting caused by *Trichoderma konongii*

8.  Compost colonized by *Trichoderma harzianum*. Note the green mould of the fungus and the red pepper mites in the corner

9.  Cap spotting caused by *Aphanocladium album*

10. Brown blotch *(Pseudomonas tolaasii)*

11. Virus disease caused by mushroom virus 4

12. Virus disease showing typical stalk elongation and small caps

13. Olive-green mould (*Chaetomium olivaceum)* and mushroom mycelium

14. Three larvae of the mushroom cecid, *Mycophila speyeri,* showing their range in size and distinctive orange colour. (Reproduced by kind permission of Dr I. J. Wyatt)

15. The life cycle of the mushroom cecid, *Heteropeza pygmaea,* showing both the paedogenetic and sexual phase of reproduction: (1) Female fly; (2) Eggs; (3) Larva; (3a) Paedogenetic 'mother' larva; (4) Pupa larva; (5) Pupa. (Reproduced by kind permission of Dr I. J. Wyatt)

16. Typical damage caused by the mite *Tarsonemus myceliophagus*. The base of the stipe is discoloured with only partial attachment to the casing.

17. Mushroom covered with jostling swarms of the red pepper mite, *Pygmephorus* spp.

18. Rosecomb

# Preface to First Edition

The mushroom crop in the United Kingdom is the most valuable single horticultural commodity, although there are fewer than 500 producers. Farms vary in size from those using 100 tonnes of compost or more every week to the small part-time producers using 5 or 10 tonnes. Because of the need to move bulk materials, the industry is mechanized and automated. There is also a considerable degree of sophistication in environmental control and in the development of new techniques, which have become the hallmark of this progressive industry. Monocropping on this scale and at this intensity is fraught with potential problems, any one of which is capable of drastically reducing crop yield or quality. The mushroom grower must always be aware of these dangers and it is very important for him to recognize the first signs of any disorder. With correct identification and the application of existing knowledge, mushroom disorders can often be successfully treated.

This book is intended to inform those interested in the mushroom industry of the causes and appearance of the most frequently encountered disorders. Most of these occur wherever the crop is grown, so the information is equally relevant for growers anywhere. By careful observation and from the descriptions of the disorders, it should be possible to make an identification and then apply the appropriate measures to combat the problem. Chemicals used for control are regularly changing, but the basic principles continue to apply. The mushroom crop is ideally suited to the use of biocontrol technology, and there have already been considerable developments in this field. It seems likely that in the future we shall see advances in the use of biological agents, which may ultimately replace pesticides as the grower's main aid in the fight against crop loss. Control of disorders is, however, often only accomplished by a combination of all the methods available. Such methods include cultural techniques, the use of resistant strains, pesticides and the often quoted but difficult to define practice of hygiene. We have attempted to integrate all these approaches and, where appropriate, to present a logical and analytical approach to the control of mushroom problems.

For those who may not be familiar with all the terms used in this book, a glossary is included on page 165; a list of books suitable for further reading will be found on page 168.

We wish to express our appreciation to our colleagues, and in particular to the members of the Agricultural Development and Advisory Service Mushroom Group. Also, to the many growers who have helped us over the years and have shared their pest and disease problems with us. We are especially grateful to Aoife O'Brien for her help when we were assembling the photographs, to Simon Rowlands for redrawing *Figures 6.2, 6.4* and *6.5* and drawing *Figure 5.2*, and also to Andy Adams who drew the *Frontispiece*.

The photographs reproduced as *Figures 1.1, 1.2, 1.4–1.11* and *3.6* were taken by RHG; those reproduced as *Figures 3.3, 3.4, 3.5, 4.1–4.11, 5.1, 6.3* and *7.1-7.3* were taken by JTF; *Figures 8.1–8.12* and *9.1-9.5* are reproduced by kind permission of the Glasshouse Crops Research Institute; *Figure 1.3* was drawn by PFW and *Figures 3.1* and *3.2* by RHG/PFW: other individual attributions appear in the captions.

July, 1986

J. T. FLETCHER
P. F. WHITE
R. H. GAZE

# Preface to Second Edition

This revision includes new information on pest and disease incidence, biology and control. Since the first edition, important pesticide legislation has been introduced in the U.K. and this is taken into account in the relevant tables and text. To make the book more useful for the recognition of problems, additional photographs have been included, some in colour.

*Figures 8.1–14, 8.17–20* and *8.22–8.28* are reproduced by kind permission of the Institute of Horticultural Research, Littlehampton; *Figures 1.3* and *8.21* were drawn by PFW and *Figures 3.1* and *3.2* by RHG/PFW; other individual attributions appear in the captions.

June, 1989
J. T. FLETCHER
P. F. WHITE
R. H. GAZE

# 1

# Mushroom Growing

## Introduction

Many readers will have an intimate knowledge of mushroom growing and the different production systems employed. A few will come to this book without such knowledge and it is mainly for them that this chapter has been included. The mushroom referred to in this book is *Agaricus bisporus*, although reference is occasionally made to *A. bitorquis*. The brief description that follows — of the principal mushroom-production processes and the different production systems and buildings that house them — is intended to act as a narrative glossary (a glossary of terms will be found on pages 165–167) and to offer some perspective to the overall environment in which mushrooms grow. For each stage and process the possible risks from contamination by pests and pathogens are highlighted so that they may be avoided.

## Culture

COMPOST

The raw material for mushroom production is straw. In the United Kingdom and many other parts of the world this is predominantly wheat straw, although straw from other grain crops such as barley is also used. Traditionally, mushroom producers obtained wheat straw as horse litter from stables. Increasingly, baled straw and poultry manure are being used as substitutes for horse litter, so that common compost mixtures would now include baled straw, horse litter and poultry manure. Horse litter may be omitted completely, thus producing a so-called 'synthetic' compost.

COMPOSTING

The process of turning these dung/straw mixtures into a suitable selective medium for mushroom production by composting (or fermentation), takes place in two distinct phases. The division between these phases is a fairly arbitrary, physical one, rather than arising from a fundamental microbiological distinction.

*Phase I composting*

The purpose of phase I composting (or simply composting) is to mix and wet the raw material and to begin the composting process during which various microflora break down the straw. With the increased use of baled straw, a pre-wetting and blending phase has become more common. During this process the raw material is made into large heaps. It is turned and wetted before being made into the long narrow stacks in which the composting process takes place (*Figure 1.1*). These stacks can be made up in the open or, more commonly now, under the protection of open-sided compost sheds. The material

**Figure 1.1** Compost stacks in a covered yard

is turned several times by mechanical compost turners (*Figure 1.2*). The phase I process takes 7-10 days. The centre of a compost stack commonly reaches 76°C (169°F), which is high enough to kill most of the pests and pathogens which may either occur naturally in the manure or straw, or are introduced via farm contamination. However, the outer layers of the stacks never reach such temperatures. Therefore, phase I compost is itself vulnerable and is also capable of contaminating later stages of production.

## Phase II composting

This can be referred to by various names, such as peak-heating, pasteurization or sweat-out. The composting process, under controlled environmental conditions, is continued in this phase until it is judged that the compost is both suitable and selective, nutritionally, for the growth of mushroom mycelium. It is primarily the control of the environment that distinguishes phase II from phase I. This 'conditioning' of the compost takes place at approximately 52°C (125°F). This is the optimal temperature which allows the microbiological activity of the microflora to render the compost both suitable and selective for mushroom mycelial growth. There is no finite point at which phase I should be translated to phase II: the more activity that

**Figure 1.2** Compost-turning machine about to turn a stack of compost

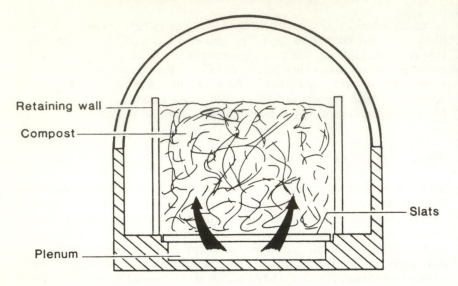

Retaining wall

Compost

Slats

Plenum

**Figure 1.3** A cross-sectional diagram of a bulk pasteurizing tunnel

takes place in one process, the less will be needed in the other. The point at which phase I compost is transported to the peak-heat or pasteurization room is dependent on complex issues of management and facilities available. It is normally more economical for phase II to last less than 1 week, thus ensuring maximum use of expensive rooms. During this process, the compost is contained in either the final growing container (shelves, trays or deep troughs), or is processed in bulk. In the latter instance, air is forced through the compost rather than around the container (*Figure 1.3*).

*Single phase composting*

If modern developments in composting continue, a single, controlled environment composting method, in which the whole of the process is carried out in bulk pasteurizing tunnels, could emerge. The concept of phase I and II would no longer apply as both parts of the process would be carried out in a single operation. The development of such a technique was derived from attempts to improve productivity and reduce costs. A major new impetus has been experienced as a result of an unexpected characteristic of the composting method, namely the absence of smell. Phase I and particularly pre-wetting can, for short periods, produce unpleasant odours. Many composters are experiencing increased 'environmental' pressure to avoid this aerial pollution. A system of composting in which the wetted blended ingredients are filled into bulk pasteurizing tunnels promises to overcome this problem.

## PASTEURIZATION

An integral part of phase II composting is the peak-heat or pasteurization process, when the temperature both in the air and in the compost is raised to 60°C (140°F) for several hours, to eliminate potential pests and pathogens. This process may take place at almost any stage in phase II composting, depending upon compost or management requirements but commonly occurs near the beginning of phase II, followed by conditioning of the compost at approximately 52°C (125°F) for the remainder of the period. Because phase II composting provides a medium selective for mushroom mycelium and incorporates a pasteurization period, failure of either function has severe implications. Once pasteurization has taken place, the compost is then vulnerable to contamination by pests and pathogens. As the severity of effect of most pests and pathogens is related both to the extent of contamination and the length of time for which it occurs, it is evident that this vulnerability is extreme.

## SPAWNING AND SPAWN-RUNNING

Once a compost has been prepared by the two phases of composting described, it is ready for spawning with the mushroom mycelium. Spawn — mushroom mycelium — is generally cultured on sterilized rye or millet grain and produced by a few specialist manufacturers.

Spawning is the process of mixing the spawn into the compost. Spawn is added to, and intimately mixed with, the compost by various mechanical means, depending on the growing system employed, at a rate of approximately 0.5% by weight (economical optimum).

Spawn-running is colonization of the compost from the grain inoculum. The environmental conditions required for a successful spawn-run are, primarily, a compost temperature of 24°C (75°F) and high relative humidity to prevent the compost from drying. Carbon dioxide levels up to 2% are beneficial and, increasingly, this environment is being achieved by re-circulating air within the spawn-running room, the air being cooled if necessary. Spawn-running, like phase II composting, may take place in the final growing containers or in bulk, in facilities similar to those described for phase II composting.

It is generally accepted that modern spawn is free from pests and pathogens because of the strict hygiene measures practised during manufacture. Once the spawn arrives on a commercial farm, however, it is vulnerable to contamination. Introduction of pests and some diseases during spawn-running can be very damaging, partly because of the early stage of production, as a generalization, the earlier that contamination takes place, the worse the effect on the crop and partly because of the vulnerability of the rapidly growing spawn and partially colonized compost. Bulk spawn-run compost is probably particularly vulnerable.

SUPPLEMENTATION

It is now quite common for compost to be nutritionally supplemented, normally at the time of spawning. Some systems of growing allow this to be done with advantage just before casing but usually it is at spawning.

Whilst the exact reasons for this nutritional boost are not fully understood, the beneficial effects on yield levels are increasingly thought to be at least cost effective, and can be as high as 20%. The supplement, usually a proprietary, high protein, slow release material, is mixed into the compost at the same time and in the same way as the spawn. The technique is not used where control of compost temperature is known to be a problem, where adequate mixing is difficult, or where there are suspicions concerning compost selectivity.

CASING

To promote sporophore production by *Agaricus bisporus,* a relatively biologically inert material, added as a surface layer to the compost, has been found to be necessary. This 'casing' layer is usually a mixture of peat and limestone 3.8-5 cm (1.5-2 inches) deep. The casing layer, which must have a neutral or alkaline pH, provides (in addition to stimulation of fruiting) anchorage for the sporophores and water-holding reserves essential for the high yields now necessary for crops to be grown economically. Casing material is easily contaminated and, if this occurs, can result in serious outbreaks of pests and diseases.

SPAWNED CASING

A technique first developed in Ireland, in which the casing layer itself is spawned, is being increasingly adopted. By mixing spawn-run compost, or specially produced casing spawns, into the casing before it is applied, the evenness and speed of case-running (see below) is greatly enhanced. Whilst the advantages of this technique are considerable, so are the implications for the spread of disease and some pests. Success of the technique, if spawn-run compost is used, depends entirely upon the selection of healthy compost, free of pests and diseases. It is difficult to ensure such selection, and the use of casing spawn may be preferable. Ordinary spawn cannot be used as it does not run in casing.

## CROPPING

Casing is usually applied when the compost has been fully colonized by mushroom mycelium, normally after approximately 2 weeks of spawn-running. There then follows a period, often called case-running, when the casing itself is colonized by mycelium. Optimal environmental conditions for the two mycelial-running phases are essentially the same.

In some growing systems, particularly shelves, it has been found to be advantageous to disturb the casing when mycelium has run about a third of the way up it. Advantage may stem from a variety of causes, but the major ones are probably a re-texturing of poorly structured casing and, more importantly, an even distribution of mycelium throughout the casing layer. In this respect, the technique is similar to spawned casing. The technique, often called ruffling, is done either by hand, with crude rake-like devices or, on modern shelf farms, with rotary, gantry equipment. Whilst not as potentially dangerous in a pest and disease context as spawned casing, the implications for the spread of pests or diseases in the casing are quite clear and require strict attention.

Once the mycelium has reached the surface of the casing, the mushroom is induced to fruit by reducing both the air temperature (to 16-18°C (60-65°F)) and the carbon dioxide concentration in the air (to 600-1000 ppm). Fruiting occurs in well-defined flushes or breaks, the first beginning about 3 weeks after casing and continuing at roughly weekly intervals. In commercial practice, four or five flushes are picked before the crop is removed to make room for the next; thus, five or six crops are often taken from each cropping room in a year. The individual flushes tend to produce progressively fewer mushrooms, the greatest weight of mushrooms normally being produced on the first flush.

Access to the crop by personnel is greatly increased during cropping. The casing is watered at regular intervals after its application and pickers move in and out of the houses to harvest the crops. Any pests or diseases which do occur usually increase in incidence during the life of the crop. Not only do pests and diseases spread within crops but there is the particular risk of transmitting contaminant pests and diseases from infected crops back to younger ones. Pest and disease identification and staff management are crucial if levels are to be controlled within reasonable bounds. The most dangerous and severely affected crops are likely to be the oldest. The methods of crop termination, compost emptying and disposal are, therefore, most pertinent to effective pest and disease control.

## Mushroom-growing systems

Phase II composting, spawn-running and cropping are three distinct processes of mushroom production, distinct in that the environmental conditions required are different. Normally, these processes are carried out in rooms designed to provide these different environmental conditions, with the compost being moved from room to room at the appropriate time; when this is the case, the system is described as a three-zone system. Not uncommonly, phase II composting may be carried out in one room and spawn-running and cropping in another, this being a two-zone system. Some growing systems are, again, different in that all three processes are carried out *in situ* in the same room; these are described as single- or one-zone systems. These manoeuvres may affect the incidence of pest and disease in several ways. The lack of compost movement in single-zone systems in many ways reduces the dangers of infection or contamination, although posing other problems such as the need to filter all houses against the introduction of virus. On the other hand, the lack of buildings designed specifically for a single process may make it difficult to achieve efficient peak-heating or composting and, perhaps, ideal spawn-running conditions.

### TRAY SYSTEMS

Trays vary considerably in size but are mostly in the range of 0.9 x 1.2 m (3 x 4 ft) to 1.2 x 2.4 m (4 x 8 ft) and are between 15 and 23 cm (6-9 in) deep. The tray system has developed into a predominantly three-zone procedure. The trays are moved by fork lift from one room to another and the processes of tray filling, spawning and casing are carried out by mechanized tray-handling lines (*Figure 1.4*). In the cropping houses, the trays are usually stacked four or five high (*Figure 1.5*). Bulk phase II composting and, in a few cases, bulk spawn-running, are being used in some tray systems. In these instances, although the system is still a three-zone procedure, compost is not filled into the trays until it enters the cropping zone. Where this practice is adopted, the trays themselves have not passed through the peak-heat and are, therefore, more likely to be contaminated.

### SHELF SYSTEMS

This is, traditionally, a single-zone system in which compost is filled into shelves, three or four high, at the beginning of the production cycle (*Figure 1.6*). Increasingly, phase II composting is carried out in bulk, thus making it a two-zone system. In Holland the system has become a full three-zone system, with both phase II composting and spawn-running being carried out in bulk. Experience of the incidence of pests and diseases in bulk spawn-running and,

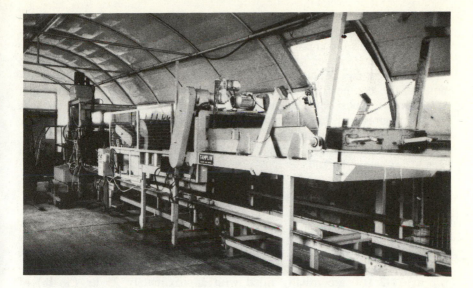

**Figure 1.4** A tray-handling line

**Figure 1.5** Mushroom trays

to a lesser extent, in bulk composting, is limited. Although bulk processing is highly efficient in many ways, it provides a new opportunity for infection. In traditional systems, compost is processed in small units and contamination is often isolated. However, where compost is handled in bulk, in quantities up to 100 tonnes, any contamination may be mixed throughout the whole volume of the compost. Additionally, circumstantial evidence indicates that spawn-run compost may be more vulnerable to certain problems, such as virus or mummy disease. Although these new methods undoubtedly offer many outstanding economic advantages, the associated potential contamination by mites, nematodes, cecids and compost-carried pathogens and moulds should not be overlooked.

BAG AND BLOCK SYSTEMS

The evolution of mushroom culture in polythene bags is complex but this system is viable when used in conjunction with bulk phase II composting. The bags, which vary in height and width and have a bed surface area of 0.1-0.2 m² (1-2 ft²), are usually filled with approximately 25 kg (56 lb) of compost (*Figure 1.7*). The system has reached its current zenith in Ireland where, in association with the socio-agricultural structure and commercial composters, it forms an expanding, competitive industry. The bags of compost may be tiered on racks but, in the simplest and best system, are placed

**Figure 1.6** A modern shelf system

**Figure 1.7** Mushrooms growing in polythene bags

on one level on the floor of the houses. In England, too, commercial composters are becoming associated with the system, supplying spawned phase II
compost, ready bagged. Thus, although the system is in fact a two-zone one,
it is, from the grower's point of view, single zone. A modern variation on the
system is one in which the phase II compost is spawned, compressed into
blocks and wrapped in plastic. The blocks are then fitted into existing trays or
shelves, or are used in a distinct system on specially made racks, tiered three
or four high (*Figure 1.8*). The divorcing of phase II composting from the
growing site generally reduces the incidence of pests and diseases, although
the previously mentioned implications of bulk phase II composting remain.

DEEP TROUGH SYSTEM

The final system extant in the British Isles is one that has evolved from the
technology of bulk processing. Bulk phase II composting and spawn-running
are carried out in a deep trough, under which runs a plenum (ventilation system), allowing air to be forced or sucked through the compost in a manner
similar to that in bulk processing tunnels. Cropping is also carried out in the
trough, thus completing the single-zone cycle (*Figure 1.9*). The system is
new and still partly developmental: its effects on the incidence of pests and
diseases are, therefore, largely unknown, although the identification of any

**Figure 1.8** Blocked spawned compost, laid out on racks

problems is greatly facilitated by the arrangement of the cropping area, which is in two or three 'table-top' surfaces in each cropping house.

## Buildings

The types of mushroom-production buildings employed in Britain are very variable. Lee Valley-type insulated plastic tunnels are increasingly common (*Figure 1.10*), and many older or larger farms use portal-frame and block-walled buildings (*Figure 1.11*). Farms recently built on the Dutch design are tall, factory-like structures made out of aluminium coated insulated panels. Almost every type of building imaginable is employed, including under-ground quarry galleries and railway tunnels. Where purpose-built houses are not used, the implications of possible inadequate composting, peak-heating and hygiene should be considered carefully. Where purpose-built houses are used, arrangement of these on a farm may have considerable bearing on pest and disease control (*see* Farm design, pages 22–24).

**Figure 1.9** The deep trough system

**Figure 1.10** Lee Valley insulated plastic growing houses

**Figure 1.11** Traditional portal-frame and blockwalled growing houses

# 2

# Disorders — Symptoms and Causes

## Introduction

Mushroom culture is a very carefully controlled biological system which aims to produce a maximum crop of perfectly formed mushrooms. The success of the system is dependent upon many interacting factors and it is not unusual for crops to achieve less than the desired production level, which in itself may be well below the maximum possible. In addition, a proportion of the mushrooms from each crop are less than perfect in shape or discoloured and the crop value consequently reduced. Biotic factors — the various pests and pathogens — often produce the most obvious symptoms, although certain abiotic factors, such as unsuitable temperatures and the availability of water, can have equally noticeable effects. In this book we are concerned chiefly with the biotic factors and the damage they cause, although we also describe a number of well-known disorders where there is no known biotic cause.

By definition, a disorder is any deviation from the normal, however small, but from the practical viewpoint, a disorder is a readily recognized abnormality resulting in a reduction in yield or quality. The expression of the symptoms of disorders in mushroom culture depends upon two major factors:

1. The stage of development of the crop;
2. The cause of the disorder.


## Symptoms in relation to crop development

The mushroom crop has well-defined stages in its development. First is the introduction of healthy mycelium (spawn) into compost, which, in normal circumstances, is followed by extensive mycelial growth until the whole of the compost is colonized. Sporophore production is induced by the application of a casing layer (usually a peat and chalk mixture) on top of the colonized compost. Once the casing layer has been applied, the mushroom mycelium undergoes a sequence of physiological and morphological changes, ultimately resulting in the growth of the fruiting bodies (sporophores) of the fungus, which we recognize as mushrooms.

On a microscopic level, the first indication of mushroom formation is the aggregation of the mushroom mycelium just under the surface of the casing. These mycelial aggregates increase in size, and are sometimes referred to as the 'spawn thickening'. The biological stimuli present in the casing which induce the production of mushroom initials are not fully understood but, on occasions, the mechanism is super-efficient and vast numbers of initials are formed. The number of sporophore initials may be so excessive that a dense, impervious layer is produced just under the casing surface, impeding the normal movement of water and gaseous exchange to the detriment of the crop.

After the formation of sporophore initials, growth and tissue differentiation occurs until the shape of the recognizable mushrooms can be seen. At an early stage in their development the undifferentiated sporophores are known as pin-heads. Differentiation is followed by a stage of rapid cell enlargement, especially in the region of the stalk (stipe) and, in particular, that part of it nearest to the cap (pileus). The developing cap produces growth-regulating materials which govern the growth of the sporophore; at this stage it is important that water uptake is not inhibited. The cell-expansion phase of sporophore growth is very rapid and, when uninhibited, results in a good quality mushroom.

It is easy to see that any disturbance in the normal development of the sporophore will result in symptoms which are recognizable to the grower. At the earliest stage — at spawning — compost that is unsuitable because of nutritional, environmental, physical or biotic factors will discourage colonization by the mycelium, thus reducing the numbers of mushrooms subsequently produced. Similarly, factors which affect mushroom sporophore development following production of initials are likely to result in incomplete development or distortion of the fruiting bodies. Very often, the symptoms are diagnostic or give a good indication of the cause of the disorder, which may then be related to the time that it first occurred. In this way, the source of the problem may be identified, e.g. wet bubble symptoms in the first flush is often an indication of contaminated casing.

## Causes of disorders

Disorders can be caused by one of, or an interaction of, a number of factors, both abiotic and biotic. For most growers it is the biotic factors that give the greatest concern, as it is the persistence of these on the farm which often affect profitability. In addition, some of the biotic causes of disorders are the easiest to identify and also to control. The symptoms may be similar, whatever the cause, so it is essential to establish the correct cause in order to rectify the problem.

BIOTIC FACTORS RESPONSIBLE FOR DISORDERS

The most commonly encountered biotic causes of disorders are insect pests, mites, nematodes, parasitic fungi, antagonistic fungi, pathogenic bacteria and viruses.

## Insect pests

Most disorders are caused by various species of Diptera (flies). The flies are often attracted to the mushroom crop in its various stages of production, and their larvae may feed directly on the mycelium, swarm over the sporophores, or tunnel into the developing or developed sporophores. The symptoms resulting from fly attack may, therefore, vary from a reduction in yield due to loss of mycelium, to discoloration and damage to the mushrooms resulting from direct attack. Tissues that have been physically damaged by flies often become colonized by bacteria which cause soft rot, thereby accentuating the problem.

## Mites

Like fly larvae, mites may feed on mushroom mycelium and on the developed mushrooms, where they can cause surface discoloration. These mites may also live on other fungi (e.g. the antagonistic fungi, *see below*) found in mushroom culture. When mites are numerous they can be a source of irritation to pickers, thus reducing efficiency. Symptoms caused by mites, therefore, include not only surface discoloration of mushrooms, but also reduced yields.

## Nematodes

These are very small and generally are visible only when they aggregate into clusters on the casing surface. The most destructive nematodes feed on the mushroom mycelium, which may show bare patches or sunken areas of bed, and the yield can be reduced significantly.

*Parasitic fungi*

Various fungi are known to be parasites of the cultivated mushroom. They are frequently recognized by their spore-producing structures, or by the symptoms shown by the affected crop. Most of the damaging fungal pathogens attack the sporophores and not the mushroom mycelium. Generally, the earlier the attack, the more distorted will be the ultimate mushrooms, for example the sclerodermoid masses resulting from infection at the initial stage by *Mycogone perniciosa*. It is possible that some fungi (for example *Diehliomyces microsporus*) may actually attack the mycelium of the crop. Symptoms of fungal attack can vary from decay or severe distortion of the sporophore to complete loss of yield.

*Antagonistic fungi*

The relationships between mushrooms and some of the fungi that can occur in mushroom crops are only partly understood. Many of these fungi become established because the physical and chemical environment of a poorly prepared compost is favourable for their development. Some moulds grow together with the mushroom mycelium and compete for nutrients; others are antagonistic and, once established, prevent the mushroom mycelium from growing into the affected compost. In both cases, the end result is yield reduction. It is also likely that some moulds produce toxins which may be volatile. These can induce distortion in the developing mushrooms. Moulds are often recognizable in the compost or on the casing surface by the coloured spores or mycelium they produce and, for this reason, have frequently been given descriptive names, such as olive green mould, plaster mould and lipstick mould.

*Pathogenic bacteria*

Some bacteria have a vital role in the successful production of mushrooms but others can cause serious disorders. Worldwide, the most common and the most investigated bacterial disorder is bacterial blotch, which discolours and sometimes disfigures the developing or even the marketed mushrooms.

There is some evidence that pathogenic bacteria can be present within apparently healthy mycelium but with no obvious effect on mycelial growth: symptoms may then develop when sporophores are produced. For instance, diseases such as drippy gill and mummy disease are caused by bacteria or similar organisms that are present within the mycelium before symptoms become apparent. Bacteria are known to cause distortion, discoloration and decay; they may also be responsible for the serious crop loss which occurs with mummy disease, although this has still to be proven.

*Viruses*

Many fungi are known to contain viruses or virus-like particles. There are relatively few instances where the presence of these is associated with disease symptoms, *Agaricus bisporus* being one of the few. Viruses in fungi differ from many other plant viruses in that they have predominantly double stranded ribonucleic acid. Also, there are no known virus vectors for fungal viruses, they are all transmitted in mycelium by anastomosis or in fungal spores. Proving that they are responsible for abnormalities in their hosts has also been difficult because it is first necessary to purify them and then introduce them into the host fungus. Exhaustive work on this aspect has yet to be completed for all the viruses found in *A. bisporus*, where a number of different particles of various shapes and sizes have been found. The precise effects of the different particle types have still to be established and there is, therefore, an element of uncertainty concerning the symptoms produced by viruses in mushrooms. What is certain, however, is that viruses can cause considerable reductions in yield. Distortion and growth abnormalities in the sporophore — giving elongated stipes, early opening of caps, and angled caps — are all symptoms that have been attributed to virus diseases; however, it is also known that these symptoms can be caused by other factors.

VIROIDS, MYCOPLASMAS AND RICKETTSIAS

These three groups are known to cause diseases in green plants and animals but have not so far been described as a cause of disease in mushrooms. They are all difficult to study and easy to overlook. It is possible that, individually, they may be playing an important role in some of the diseases at present attributed to other causes, e.g. viroids in virus diseases and rickettsias or mycoplasmas in mummy disease.

ABIOTIC FACTORS RESPONSIBLE FOR DISORDERS

The main abiotic factors are temperature, relative humidity, carbon dioxide concentration in the air, an excessive moisture level in the compost and casing, and the presence of toxic chemicals in the atmosphere, compost or casing. Either alone or in combination, these factors may, where they deviate from the optimum, result in poor mycelial growth — and, therefore, poor cropping — or distortion of the developing or developed sporophore to give symptoms such as hard gill, rosecomb or mass pinning.

# 3

# Effective Pest and Disease Control

## Introduction

The successful control of the many pests and diseases that may occur on any mushroom farm requires a complete farm strategy. Often, the control of a particular pathogen or pest will depend upon several factors, which must all be taken into account (*Table 3.1*). It is also essential to understand the biology of the organisms: for instance, it is important to know their method of entry to the crops and their means of spread within and between farms. This, together with the chemical control measures available, will give the greatest chance of achieving a satisfactory level of control.

The most important aspects listed in *Table 3.1* are examined in detail in this chapter in an attempt to integrate these into a whole programme which, with minor modification, would be applicable to any farm.

## Farm hygiene

The importance of hygiene in the successful control of most diseases and many pests is now universally recognized; what is perhaps less commonly realized is the actual degree of its importance. Hygiene is, and is likely to remain, a major method of pest and disease control: it is the foundation upon which the success of all other control techniques depends.

After a farm has been in production for some time, the major sources of pests, pathogens or weed moulds tend to be from the farm itself. The objectives of any hygiene programme must be:

1. The exclusion of pests and diseases from the production cycle;

2.  The elimination of pests or pathogens, or at least the containment of them, if and when they occur during cropping;
3.  The destruction and elimination of pests and diseases present within a crop at its termination.

Such measures should help to reduce the overall contamination level and to ensure a clean start for buildings and compost containers for subsequent crops.

Farm design, the use of disinfectants, filtration, peak-heating, cook-out and techniques of isolation and containment, are all the building blocks of a successful programme. As in any form of construction, the constituent parts must be arranged in an orderly and logical fashion. The omission or weakness of any part risks the whole.

**Table 3.1**   Factors influencing the incidence of pests and diseases on a mushroom farm

**Farm hygiene**
Farm design
Peak-heat
Filtration
Cook-out
Management
   Staff
   Isolation of disease
Disinfectants
**Farm hygiene programmes**
**Chemical control**
   Choice of pesticides
   Methods of application
   Phytotoxicity
   Pesticide resistance
**Environmental control**
**Resistant strains**

FARM DESIGN

The farm design is vitally important for the successful control of pests and pathogens. It must, of course, be accepted that the design of any farm, once set, is difficult to change. The form which it takes is often governed by factors not directly influenced by consideration of control procedures: restrictions of site size and shape, the evolution of the farm over an extended period, and the logistic needs of compost handling, usually dictate a plan which produces compromises. A good design for logistic purposes goes a considerable way towards fulfilling the requirements for the successful control of biotic disorders. The basic aim in design, from a pest and disease control point of view, must be to prevent flow of materials, personnel and debris from areas of the maximum likely pest and disease contamination to clean areas. By examining the areas of maximum risk in graphic form on a stylized farm plan, a complex situation can be made relatively clear. The same exercise, carried out with an actual farm plan, highlights areas where extra precautions in terms of filtration, management of staff or the use of disinfectants may be necessary. Examples of the value of such an analysis are illustrated in the following hypothetical farm designs.

## Design 1

This plan (*Figure 3.1*) shows the major areas of vulnerability; the processing line, peak-heat, spawn-running rooms, casing store and mixing area. It also shows the major source and accumulation of pests and pathogens — which is in the cropping area — and the minor source, particularly of pests, which is in the composting area. These facilities are best arranged so that crossing in the flow of materials is avoided and the high-risk and vulnerable areas are well separated. With this design, filters, disinfectants and the management of staff movement should allow the best chance of effective pest and disease control.

## Design 2

This example (*Figure 3.2*) demonstrates a not uncommon situation where piecemeal development and site restrictions do not allow clear separation of highly contaminated areas and areas of high vulnerability. In such a situation it is essential to:

1. Provide filtration of peak-heat and spawn-running rooms;
2. Prevent personnel movement of all sorts across the common area, or at least make sure it is strictly and logically controlled;
3. Use a wide spectrum of disinfectants, in the common area frequently;
4. Fit dust filters to the air inlets of cropping houses;
5. Take care to protect the casing store of which is dangerously near the cook-out and box-emptying areas (*Figure 3.3*);
6. Scrupulously clean the processing line between operations as it is too close to the compost yard and the route taken by the spent compost.

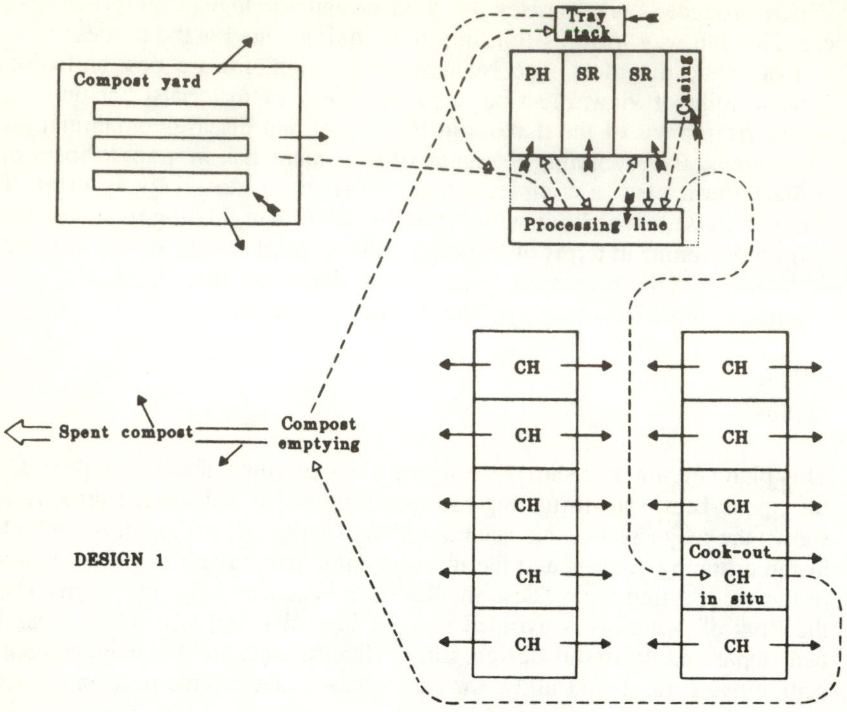


**Figure 3.1** A farm plan demonstrating an arrangement of facilities favourable to good hygiene. Casing = casing store + mixing area; CH = cropping house; PH = peak-heat room; SR = spawn-running room; -----▷ the route of one crop; ——► sites of contamination; ■—► vulnerable points

These points hold good for any farm, whatever the design. In this instance, two hypothetical tray farms have been used. Shelf farms will bring different problems to light. Filling and emptying operations are dirty and inevitably carried out in close proximity to other cropping houses. Both shelf and phase II systems; bags and blocks; share the problems of two-zone systems in that the vulnerable spawn-running phase is carried out in a cropping house amidst other cropping houses. The extra protection given to only spawn-running rooms on three-zone tray farms has, therefore, to be extended in these circumstances to all cropping houses. Whatever the system of growing, analysis of farm layout can illustrate shortcomings which may be overcome by long-term capital works, or by the institution of extra hygiene measures.

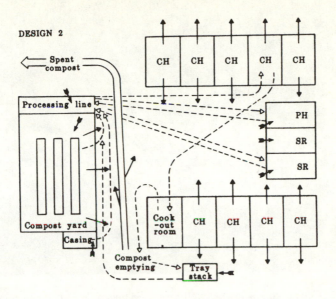

**Figure 3.2**   A farm plan illustrating many aspects of design unfavourable to good hygiene. Abbreviations and symbols as in Figure 3.1

PEAK-HEAT

Peak-heating of compost is now a standard procedure, and there is a risk of underestimating its value in ensuring that the compost emerging from the 'peak-heat' rooms is (to all intents and purposes) pest and disease free.

As compost densities in trays and shelves have risen, so lower air temperatures are required to ensure that the desired compost temperatures are achieved during phase II and that overheating is avoided. Little evidence is available but it seems logical to aim to achieve an air temperature of 60°C (140°F) for at least 1 hour, to ensure the death of pests and pathogens on and in the surface layers of compost. Every effort should be made to prevent wetting of upper trays by condensation because this may upset the composting process and reduce the effiency of the peak-heat.

Bulk processing, in that it also incorporates a 'peak-heat', fulfils the same function. In this instance, the compost most likely to escape the process and provide a source of contamination will be that along poorly insulated walls or in dead spots produced by uneven filling. Because of the bulk nature of the process, large quantities of compost can become completely contaminated, even from such small untreated areas.

**Figure 3.3**  Casing store, showing well-protected materials and clean equipment

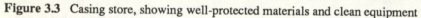

Another aspect of bulk processing that should not be forgotten is that trays or shelves used as the growing containers have not been heated to 60°C (140°F) during phase II. They must, therefore, be heat-treated or cleaned with disinfectants.

FILTRATION

Virus diseases have made filtration of peak-heat rooms and spawn-running rooms virtually essential if periodic, debilitating attacks of these diseases are to be avoided.

Affected crops produce a 'fog' of mushroom spores carrying virus particles all over the farm and from which vulnerable stages of the crop must be protected. In this instance, therefore, it is the spores of the mushrooms themselves that must be excluded and, consequently, filtration must be designed for this purpose.

The normal recommended level of filtration is 98% efficiency using a 2 μm mesh filter. Such filters are always provided with prefilters to collect coarse material. Prefilters should be changed frequently. The duration of the life of prefilters can be judged only by experience, taking into account airflow measurements, to ascertain when clogged prefilters begin to impede airflow into the houses.

Similarly, when filters are first fitted to peak-heat rooms or spawn-running rooms, their effect upon the air distribution systems must be carefully calculated. Even clean, newly installed filters cause an increase in air resistance and can, therefore, considerably affect the efficiency of fans already installed.

It should be remembered that, for filters to work efficiently, the ducting must be absolutely airtight. Many filtration failures can be attributed to poorly installed or maintained ducting, which allows air to bypass the filter banks and, thus, to deliver a damaging load of virus-carrying spores on to compost at cooldown or during spawn-running. The need to filter spawn-running rooms is now often obviated by recirculated cooled-air systems.

All that has been said concerning the necessity to filter peak-heat rooms must also be applied, with even more emphasis, to bulk processing rooms, where the problems that occur if virus enter at that stage are greatly magnified. Whether or not to filter at other stages of production is more debatable. There is some recently available evidence to suggest the need for exclusion of specific weed mould spores from bulk pasteurizing and bulk spawn-running tunnels.

Infected mushroom spores do not pose a great threat to crops once the casing has been applied. In some circumstances, the spores of fungal pathogens may be carried through the air inlets into cropping houses. If it is suspected that this is a significant factor in the introduction or spread of such pathogens, dust filters can be used.

In analysing the need for filtration of inlet ducts, the objective must constantly be borne in mind: this is to exclude pathogens, predominantly viruses and, to a lesser extent, *Verticillium* (*see* page 52), from the sensitive stage of production during composting and mushroom cropping. The situation already described refers exclusively to farms with the three-zone system (*see* Chapter 1, pages 8–12) with the crop grown predominantly in trays.

The pattern of need for filtration becomes quite different in two- or one-zone systems, where one or more of the vulnerable production processes are carried out in what is, ostensibly, a cropping house. In these circumstances, 2 μm filtration may have to be used to protect all houses.

In the past, there has been some discussion about the desirability of filtering exhaust ducts from cropping houses, to ensure trapping of potential virus-infected mushroom spores before they reach the rest of the farm. This is difficult to do in practice and, except in extreme circumstances where a continual production of open mushrooms is contemplated, probably is not essential if there is effective filtration of the processing rooms.

COOK-OUT

The most heavily contaminated area on most farms is where the older crops are about to be terminated. Elimination of all the pests and diseases that have built up within these crops and on the structures, shelves and boxes is one of the essential steps in any effective programme.

There are two effective treatment options, steam or methyl bromide fumigation. In practice, a cook-out can vary from the ideal which is treatment of the terminated crop *in situ* to treatment in specialized buildings or, least effective of all, treatment of the boxes after they have been emptied. Because of physical restrictions, poorly insulated or sealed buildings, or fear of damage to structures or trays, the ideal is often not practised.

*Heat*

If it is possible to cook-out *in situ* with steam, a temperature of 71°C (160°F) held for 2 hours will effectively kill all pests and pathogens. Most work on thermal death-points indicate that this temperature is adequate, even for *Diehliomyces microsporus* (false truffle, page 64), which is generally recognized as being one of the most difficult to control. It should be stressed that all the compost, casing and growing containers, and not just the air, should reach this temperature; recording probes are essential in order to ensure that this has been achieved.

*Methyl bromide*

Fumigation with methyl bromide as an alternative to steam finds less favour now than in the past, although it is still used (*Figure 3.4*). Because of its highly poisonous nature, fumigation with methyl bromide can be done in the United Kingdom only by contractors who are familiar with such factors as the CTP (concentration x time, product), conditions of application and leakage problems. The advantages of methyl bromide are that it does not damage either the structures or the growing containers; on the other hand, it is a highly dangerous material and is less effective than heat against some fungi and bacteria.

*Other chemicals*

If neither heat nor methly bromide can be used *in situ*, some form of surface sterilization should be attempted before the crop is moved. Fumigation with formaldehyde is undoubtedly quite effective, using 283 ml (0.5 pint) of

**Figure 3.4**   House sealed for methyl bromide fumigation

formalin per 28 m³ (1000 ft³ ), vaporized with steam. Alternatively, the house can be sprayed with formalin (2% plus wetter) or one of the other disinfectants. The aim of these chemical treatments is primarily to sterilize the surface of the trays and the floors of the houses. The walls and ceilings must also be treated, although these may be less heavily contaminated.

The choice, then, of a steam cook-out or methyl bromide fumigation, either *in situ* or on the trays only, is influenced by many factors, not least of which is the physical ability to carry out the process successfully. The type of pest or pathogen most prevalent on the farm is another consideration. It is important that pests such as nematodes, cecids and mites, and pathogens such as those causing false truffle or virus diseases, are effectively eliminated during cook-out as this is the one upportunity during the cropping cycle when they can be controlled. Whatever the choice of method, it is good advice to have as effective a cook-out as possible, despite the superficially high cost and, at times, apparent lack of need to do so.

MANAGEMENT

*Staff*

Effective staff ensure the success of all operations and, in particular, the control of pests and diseases. On large farms, it is often worth employing staff with sole responsibility for pest and disease control. These should be well trained in all aspects of the recognition of disorders, so that they are able to identify problems as soon as they arise. Procedures for the removal of affected mushrooms at regular and frequent intervals, and always before watering and picking, are now well established. It is important that staff should know which diseases are highly contagious, and should use disinfectants wisely in order to prevent the spread of pathogens and pests from infested to uninfested crops. With this in mind, it is normal for work to begin at the clean end of the cropping cycle, and to work towards the oldest and usually most contaminated crop. This sequence should not, of course, be sacrosanct as the occurrence of diseases or pests in a relatively young crop should immediately relegate it down the sequence of crops to be worked.

Clothes are a major source of spread of pests and diseases and it is vital that overalls are changed regularly, preferably every day. Contaminated overalls are effectively freed from all pests and pathogens by the normal process of washing in hot water with a detergent. Similarly, pickers' knives can spread pathogens and it is important that knives and other tools that are used in crops should be regularly cleaned by immersion in a disinfectant.

*Isolation of pests and diseases*

One aspect of management that deserves particular attention is the isolation of an affected crop. This is especially relevant for the control of diseases. Once established in a crop, the diseased mushrooms themselves are the greatest source of inoculum for further spread on the farm. Early identification and isolation of these is, therefore, essential (*Figure 3.5*). Whereas almost every mushroom grower would, in theory, agree about the importance of such a procedure, in practice it is often the difficulty of carrying through a programme effectively that allows the build-up of disease.

DISINFECTANTS

Disinfectants are an integral part of any programme, being instrumental in the exclusion, containment and often (finally) the elimination of pests and disease (*Figure 3.6*). Most disinfectants are poisonous and should be used strictly according to the manufacturer's or supplier's recommendations. Recent pesticide

**Figure 3.5**  Area of disease isolated with salt

legislation in the United Kingdom has been set up to control the range of
pesticides and their application rates and methods. This legislation does not
apply to the use of disinfectants. The distinction between a disinfectant and a
pesticide is that the former will never, directly or residually, come in contact
with the crop. Most chemicals are clearly one thing or the other but it should
be remembered that if compost, casing or growing containers are treated with
disinfectants, these chemicals may then be considered to be pesticides. In
these circumstances, they will usually not have been cleared and will, therefore, have been illegally used. Similarly, if disinfectants are sprayed or misted
near compost or growing crops, they may similarly become, in effect, pesticides.

**Figure 3.6**   Foot dip containing disinfectant, at the entrance to a cropping house

Because disinfectants are not covered by the pesticide legislation, the aspects of operator safety and potential pollution must be the more closely supervised.

The effective use of disinfectants on the farm is greatly affected by the physical conditions. For instance, many disinfectants are inactivated in the presence of organic matter such as casing or compost. It is, therefore, very important to be able to wash down, as well as disinfect, all the important areas of the farm, in order to remove such material. This means that most surfaces must be concreted and the surface of the concrete finished in such a way that it is smooth and without cracks and holes, which may harbour debris. Similarly, buildings and machinery must be effectively cleaned at regular intervals and power washing, either with or without a disinfectant, must be a routine practice. It is, of course, important to ensure that all waste disinfectant is carefully ducted away and not allowed to pollute water-courses or adjoining land, where it might be harmful.

Many disinfectants are strong biocides capable of killing virtually all organisms. These can be unpleasant to use and they may be too persistent or volatile for use in enclosed spaces, particularly those areas leading to the cropping houses, or in the houses themselves. The choice of materials may, therefore, be a compromise, even before any consideration is given to the effectiveness of the material against a specific pest or pathogen. The choice of disinfectants should be limited to those tested by the manufacturers in the specific context of mushroom growing. In the open air and on large surfaces, the more general and often cheaper biocides are probably quite suitable. In these instances in particular, thought must be given to the increasing problems of pollution of water-courses, as mentioned above.

The choice of suitable materials must also be governed by the pest or pathogen to be controlled (*Table 3.2*). Most disinfectants are general biocides but some are particularly effective against certain groups of organisms. Almost all disinfectants are toxic to mushroom mycelium or sporophores and must always be used with care near the crop. In order to be effective, a disinfectant must wet the whole of the surface area to be treated. Sometimes, when dry surfaces are being treated, or mycelium is adhering to wood, air bubbles are trapped against the surface, preventing contact between the disinfectant and the surface. For this reason, a wetter may be added to the disinfectant, but some are formulated with added wetter. The manufacturer of the disinfectant should always be consulted before a wetter is added. Generally, anionic wetters are most suitable and the non-ionic types, commonly used in horticulture, are unsuitable. The most readily available anionic wetting agents are contained in washing-up liquid and front-loader washing-machine powders.

Some disinfectants need to be stored carefully to maintain their activity. This is particularly true of the chlorine-producing chemicals, which are best stored in the dark and at low temperature (ideally, 3–5°C). Stored in this way, chlorine-producing chemicals will retain their effectiveness for several months. Most commercial preparations of sodium hypochlorite contain about 10% active chlorine. The chlorine content of the concentrate can easily be checked.

CHECKING THE CHLORINE CONTENT OF A HYPOCHLORITE DISINFECTANT

Because the shelf-life of hypochlorite disinfectants is short, especially if they are not stored in a cold, dark place, it is important to check the available chlorine content at intervals, especially if large containers are bought and not used quickly.

*Procedure*

1. Prepare solution A by dissolving 12.5 g potassium iodide in a 12.5% v/v solution of glacial acetic acid in water.
2. Prepare solution B by making a 0.28 M solution of sodium thiosulphate (this is done by taking one ampoule of Convol (0.1 M sodium thiosulphate) available from BDH Limited, catalogue no. 180445C, and making it up to 357 ml using de-ionized water).

**Table 3.2** Some disinfectants used in mushroom growing

| Trade name | Chemical name (and manufacturer) | Label recommendations | Comment |
|---|---|---|---|
| Cryptogil | Sodium pentachlorophenoxide (Rhone-Poulenc) | Tray dipping, 2% solution 1 kg to 50 litres (2 lb in 10 gallons). Dipping time at least 4 seconds. Washing down, 1% solution although contamination risk makes this treatment risky. | A general biocide which prevents mushroom mycelium from growing into treated wood; a good tray-dipping material |
| Diversol BX | Chlorinated trisodium phosphate-brominated (Diversey Ltd.) | 1. Post-cropping before emptying and in empty cropping house. Fogging 170 g/4.5 litres (6 oz/gal). 2. Post-casing before pinning. Fogging 113 g/4.5 litres (4 oz/gal). 3. Spawning to final flush. Fogging 57 g/4.5 litres (2 oz/gal). | A general biocide with some antiviral properties. |
| Environ | Phenolic derivatives (Ceva Ltd.) | Pre-cropping and post-cropping, all surfaces. During cropping, walls and floors only. Also used for cleaning equipment. 0.4% solution. Footbaths 1.5% solution. Fogging: empty houses. | A general biocide with a supposedly pleasant smell; phenolics can induce rosecomb so should not be used in a developing crop. |
| Prophyl | Phenolic derivatives (I.P.P. Ltd.) | Disinfecting empty houses, equipment, concrete surfaces, crop termination, foot dips. 0.4% solution, 1 litre in 250 litres of water per 1250 m². 0.2% for concrete and 0.5% for foot dips. | Wide spectrum of activity and pleasant to use. |
| DCT concentrate | Phenolic derivatives (ERA Hygiene) | Empty buildings, walls and floors, foot baths. Fogging into ventilation ducting. 0.4% solution, 5 litres to 600-750 ft². | Wide spectrum of activity with little odour. |

| Product | Active ingredient | Rate / use | Remarks |
|---|---|---|---|
| Formalin | Formaldehyde (Various) | 5% of commercial general product, i.e. 5 litres in 95 litres washing-down water (1 gal in 19 gal water). Fumigant 0.5 litres/28 m³ (1 pint/1000 ft³) with 125 g (4 oz) of potassium permanganate. | A highly effective biocide but unpleasant to use. |
| Various | Sodium and calcium hypochlorite | For washing down and general cleaning. Approximately 1 litre in 450 litres (2 pints in 100 gal). | An effective biocide which is rapidly inactivated as soon as it contacts organic material. Concentrate must be stored in the dark at low temperatures. |
| Nuodex 87 | Dodecylamine salicylate (Durham Chemicals Ltd.) | Washing down and generally cleaning surfaces; 1-5% solution. | General biocide. |
| Panacide-M | 4,4'-dichloro-2,2' methylenediphenol (common name dichlorophen) (BDH Chemicals Ltd.) | For tray dipping, foot baths and washing down at 1:60, 1:500, 1:200 respectively. (1:200 for cleaning machinery, concrete, surfaces). | A safe-to-use general biocide which is also known to be effective against algae. |
| Sudol | Phenolic distillate (Tenneco Organics Ltd.) | Surfaces and machinery wash down at 2-3% and as a foot dip at 4%. | The only disinfectant recommended for cecid control; may cause rosecomb if in contact with developing crop. |
| Tego 51 | Ampholytic surfactants (T. H. Goldschmidt Ltd.) | General disinfectant 1%; isolation technique for Verticillium 1-2%. | Safe to use and odourless; frequently used when removing diseased mushrooms from beds. |

3. Measure 80 ml of solution A into a container, such as a conical flask.
4. Add 1 ml of the test hypochlorite solution.
5. Add solution B by 1 ml lots, while shaking the flask, until the brown colour disappears.

For every 1 ml of solution B used to remove the brown colour, the test solution contains 1% of available chlorine: i.e., to be satisfactory, 10 ml of solution B should have been used to indicate 10% available chlorine.

If the concentration of available chlorine is less than 10%, this must be taken into account when using the hypochlorite disinfectant.

## Farm hygiene programmes

Farm hygiene programmes must be implemented in all the phases of mushroom production, bearing in mind the vulnerability of the crop to particular pests and diseases at different stages. *Table 3.3* summarizes the aspects of pest and disease control outlined in the text. Although these examples of hygiene measures are not intended to serve as a blueprint for a programme, they illustrate how the various aspects of hygiene are woven together to form a complete farm programme, based upon the control requirements for different types of pests and diseases.

## Chemical control

### CHOICE OF PESTICIDES

There are a limited number of pesticides (all chemicals applied to compost, casing or the crop to control pest and disease organisms) registered for use on mushrooms. Reference has been made to the legislation current within the U.K. in the section dealing with disinfectants. Previous voluntary regulations are now legally enforceable as a result of this legislation. Some of the chemicals available for use are listed in *Tables 4.1* and *8.1* (pages 46 and 128 respectively). An authoritative and detailed statement of the position relating to cleared chemicals can be found in the annually produced 'U.K. Pesticide Guide', produced jointly by C.A.B. International and the British Crop Protection Council. It is vitally important to follow the instructions on the label, as underdosing may result in loss of efficiency and overdosing may cause crop damage. Detailed recommendations for all pesticides used on mushroom crops are given in the following chapters.

METHODS OF APPLICATION

There are various methods of application of pesticides to mushroom crops. It is important that the best method is selected with regard to the pesticide and disorder concerned. The wrong choice may result in an ineffective control or, in some cases, may cause phytotoxicity.

*Incorporation in compost*

It is difficult to mix a small amount of pesticide into a large volume of compost yet thorough incorporation is necessary to ensure adequate treatment. Liquid applications may be more effectively mixed, especially when sprayed on to the compost as it passes fixed spray nozzles. If granules are to be applied, the active ingredient is often diluted with an inert carrier to bulk up the relatively small quantity of chemical and aid even distribution.

*Casing incorporation*

The problems of efficient mixing of pesticides with the casing are the same as those with compost but can be solved by the use of modern machinery. In most instances, the only effective method is to dissolve, dilute or suspend the pesticide in the water used to wet the casing.

*Sprays or drenches*

Most chemicals are applied as a spray or drench to the casing material. To obtain an effective cover and some penetration, it is generally necessary to apply 200 litres to each 100 m² of casing surface (approximately 40 gallons per 1000 ft²). Methodical application to ensure even treatment of the whole bed surface is the main aim. It is important to apply the prescribed volume per given area, thus avoiding overdosing and possible phytotoxicity. Different spray nozzles may be needed to give even applications and correct dosage with different pesticides.

*Fogs, dusts, smokes and vapours*

With these methods of application it is essential to calculate accurately the volume of the treated house in order to calculate the correct quantity of pesticide to be applied. The distribution is often uneven. Fogs or dusts of fast-acting insecticides are sometimes used for cosmetic effect or to reduce

**Table 3.3** Examples of hygiene measures for some types of pests and pathogens

| Crop stage | Verticillium | Virus | Cecids | Complete programme |
|---|---|---|---|---|
| Compost | | *Aim* Protect from mushroom spores *Action* FARM DESIGN: site the yard away from exhaust air COOK-OUT: reduce contamination MANAGEMENT: disposal of spent compost | | *Aim* Protect from mushroom spores *Action* FARM DESIGN: yard siting COOK-OUT: site contamination MANAGEMENT: compost disposal |
| Peak-heat | | *Aim 1* Prevent carryover on trays* or shelves *Action* COOK-OUT FARM DESIGN: management, tray storage DISINFECTANTS: dip or treat growing containers as sterilization and to treat wood for cleaner emptying *Aim 2* Prevent ingress of mushroom spores *Action* FILTRATION *Aim 3* Eliminate mushroom spores or mycelium | *Aim 1* Prevent carryover of larvae on and in trays and shelves* *Action* COOK-OUT DISINFECTANTS MANAGEMENT *Aim 2* Eliminate larvae in compost *Action* PEAK-HEAT: attention to prescribed temperatures | *Aim 1* Prevent carryover on trays *Action* COOK-OUT DISINFECTANTS: tray treatments FARM DESIGN: tray storage MANAGEMENT: tray storage *Aim 2* Prevent ingress of mushroom spores *Action* FILTRATION *Aim 3* Eliminate pests and pathogens in compost |

| Stage | | PEAK-HEAT: attention to temperatures | | PEAK-HEAT: prescribed temperature regimes |
|---|---|---|---|---|
| **Spawn-running** | *Aim* Protect from pathogen spore contamination. *Action* FARM DESIGN: siting and design of casing store. DISINFECTANTS: cleansing floors, mixing machinery, peat and chalk external packing materials. MANAGEMENT: reduce dust, cropping house exhaust air emission during vulnerable periods, and careful material handling of ingredients | *Aim 1* Prevent ingress of mushroom spores*. *Action* FILTRATION: or recirculated cooled air. DISINFECTANTS: clean machinery and working surfaces. MANAGEMENT: protect compost from contamination by reducing spore emission and dust levels during handling; store spawn in clean areas | *Aim* Protect from contamination. *Action* DISINFECTANT: clean machinery and working areas. MANAGEMENT: order of working; Never proceed from contaminated crops to vulnerable processes | *Aim 1* Prevent ingress of mushroom spores. *Action* FILTRATION. *Aim 2* Prevent compost contamination. DISINFECTANTS. MANAGEMENT. *Aim 3* Protection of spawn. *Action* MANAGEMENT: clean spawn storage |
| **Casing** | | *Aim* Protect from contamination, mushroom spores and mycelial debris. *Action* FARM DESIGN. DISINFECTANTS. MANAGEMENT | *Aim* Protect from larval contamination*. *Action* FARM DESIGN: siting and design of casing store. DISINFECTANTS: cleansing of handling machinery and work surfaces; in extreme cases pasteurize casing | *Aim* Protect casing from pest or pathogen contamination*. *Action* FARM DESIGN: casing store. DISINFECTANTS: machinery working areas. MANAGEMENT: reduce contamination levels whilst casing mixed and applied. In extreme cases pasteurize casing before use |

| Cropping | | | |
|---|---|---|---|
| *Aim 1* Protect crop from spore contamination | *Aim* Reduce mushroom spore* production | *Aim* Isolation | *Aim 1* Protect from aerial contamination |
| *Action* FILTRATION: dust filters DISINFECTANTS: enhance containment methods, treat floors and foot dips | *Action* MANAGEMENT: pick closed mushrooms FILTRATION: if open mushrooms grown by choice consider exhaust filtration | *Action* MANAGEMENT: if pest is identified every effort should be made to contain it to the affected house | *Action* FILTRATION: dust filters |
| *Aim 2* Identify disease* | | MANAGEMENT: protective clothing used only in affected house; work schedules arranged to place infested houses last in order of the day | *Aim 2* Identify isolate and contain pest or disease in affected houses |
| *Action* MANAGEMENT: lighting; constant examination | | | *Action* MANAGEMENT: isolation techniques for specific diseases, maintenance of work order schedules and use of protective clothes; control vectors |
| *Aim 3* Isolation and containment* of disease | | | DISINFECTANTS: treat floors, supply foot dips |
| *Action* MANAGEMENT: cover, salt or remove before picking or watering | | | FILTRATION: exhaust filters if open mushrooms grown |
| MANAGEMENT: work only from new houses (or clean) to old (or affected); individual clothing for each house | | | COOK-OUT: employed if containment impossible |
| MANAGEMENT: fly control | | | |

COOK-OUT:
terminate very heavily
diseased crops early.

**Post-cropping**

*Aim*
Prevent disease carryover
and site contamination
*Action*
COOK-OUT/
DISINFECTANTS:
terminate *in situ* wherever
possible; failing this,
surface sterilization before
emptying or cook-out
elsewhere
MANAGEMENT: remove
spent compost from farm;
reduce dust and
contamination to newer
crops whilst so doing

*Aim*
Prevent disease carryover*
and site contamination
*Action*
COOK-OUT:
*in situ* if at all possible;
cook-out in growing
containers as an alternative
DISINFECTANTS:
treat trays or shelves
to further sterilize and
enhance physical cleaning;
cleansing of air ducts
MANAGEMENT:
remove spent compost
from the farm immediately

*Aim*
Prevent pest carryover
and site contamination
*Action*
COOK-OUT:
*in situ* if possible;
it is particularly
important to sterilize
the trays or shelves
DISINFECTANTS:
wash out cropping houses,
machinery and work surfaces
with an effective disinfectant
MANAGEMENT:
remove spent compost
from the farm immediately

*Aim*
Prevent pest and disease
carryover and site
contamination
*Action*
COOK-OUT:
*in situ* if possible plus tray
treatment if appropriate
DISINFECTANTS:
surface sterilization if *in
situ* cook-out is not
possible, cleansing house,
ducts, machinery and work
surfaces; treatment of trays
before re-use
MANAGEMENT: disposal
of spent compost; prevent
contamination of growing
crops and vulnerable
processing areas during
emptying

Measures in capital letters indicate aspects dealt with individually in the text.
*High-priority areas for specific pest or pathogen.

the spread of *Verticillium* by mushroom flies. Another often-used application method is the use of dichlorvos, vaporized in spawn-running rooms to prevent incursions of phorids.

Smokes can be used to prevent phorid flies from laying their eggs in the compost during spawn-running. However, smokes or fogs which last only a few hours are not very effective when used once or twice a week.

PHYTOTOXICITY

Most instances of obvious phytotoxicity of pesticides have occurred when application has, in some way, been contrary to recommended practice, for example:

1. When a pesticide has been mixed with casing rather than applied as a drench on the bed;
2. When the recommended volume of water used per given area to apply the pesticide has been exceeded;
3. When the actual concentration of the pesticide has been exceeded;
4. When chemicals have been used at times other than those recommended, e.g. during cropping, when the label stipulates before cropping only.

Very occasionally, unforeseen instances of phytotoxicity occur, even when the pesticide has been applied correctly according to the label recommendations. In such uncommon instances, simultaneous combinations of several factors are often present in order to induce phytotoxicity. For instance, diazinon incorporated into compost at the label recommended rate has, on occasion, caused phytotoxic effects if a sensitive spawn has been used and the spawn and insecticide were in intimate contact during application.

PESTICIDE RESISTANCE

Resistance of *Verticillium* to benzimidazole fungicides, such as benomyl (*see Table 4.1*, page 46), is now well known, as is the resistance of sciarid larvae to organophosphorous compounds, such as diazinon (*see Table 8.1*, page 128). Avoidance of such resistance should be an intrinsic part of any control strategy.

Repeated and regular applications of any chemical greatly increase the chance of resistance developing. If other, equally effective pesticides with a different mode of action are available, it is helpful to alternate them. However, in mushroom production there are rarely suitable alternatives. The answer must be, therefore, to use pesticides sparingly and never on a routine basis. If resistance begins to develop, it is advantageous to recognize the fact before pest or disease incidence has risen to levels difficult to combat by alternative methods. If pest or disease levels begin to rise, the possibility of resistance should be considered, together with other factors such as inappropriate application rates, ineffective methods of application, or lapses of hygiene.

## Environmental control

There are, unfortunately, few pests or diseases that can be controlled satisfactorily by environmental manipulation alone. The control of bacterial blotch (*see* page 76) is, perhaps, the best-documented example of success by such means, and is based on the knowledge that, for multiplication, the bacteria (*Pseudomonas tolaasii*) require moist conditions and that discoloration of mushrooms occurs only when the bacterial population is very large. By environmental control, it is possible to reduce to a minimum the time during which mushroom caps are wet. This form of disease control is quite complex, involving careful management of watering, the air temperature, and air movement. It may, at times, be difficult to reconcile the practices conducive to the control of blotch with those required to promote maximum yields. However, the benefits in reducing the quantity of third-quality, or even of unmarketable mushrooms, can be considerable. Disease control by environmental manipulation is most applicable when high ambient temperatures and humidities make it difficult to ensure good evaporation from the beds and mushroom surfaces. Fluctuations of temperature result in the frequent occurrence of the dew point, which keeps the mushrooms wet and may give rise to severe blotch symptoms. Very good temperature control and dehumidification, by means of coolers operated in conjunction with the heating system, are both essential to avoid such conditions, especially in the autumn months. It is possible that some of the other bacterial diseases may be reduced by similar environmental control.

Environmental control can also be very important in the control of false truffle (*see* page 65). Hygiene goes much of the way to controlling the problem by preventing its occurrence but there is no known chemical control; it is, therefore, difficult to eradicate the organism, once established, without manipulation of the environment. Here, temperature control of the compost is very important, in order to exploit the difference in the temperature requirement for the mycelium of the mushroom (24°C; 75°F) and for the development of false truffle (29°C; 85°F). If contamination with false truffle is known on the farm and can be expected to occur, temperatures in the compost during spawn-running should be reduced as much as possible. Such a reduction is more unfavourable for the pathogen than it is for the mushroom mycelium. Thus, a 13-day spawn-run at 21°C (70°F) will greatly disadvantage the pathogen, while having little effect on the mushroom. This form of control not only alleviates the problems of yield reduction caused by the pathogen but will, in addition to the hygiene measures practised, slowly reduce the overall site contamination. The environment to be avoided at all costs when this organism is present is a high spawn-running temperature, i.e. above 24°C (75°F). Recent experience has shown the importance of preventing excessive temperature rises, even for very short periods. There is evidence to suggest that a few hours is enough to 'activate' any false truffle spores present in the compost. Many growers are reluctant to engage in control of this problem by means of temperature control since the advent of hybrid strains of mushroom. Whilst this is understandable with reference to the need to fully colonize compost with hybrids, it remains essential in the control of this debilitating disease. The answer surely lies in the exact and positive control of temperature. Damagingly sub-optimal temperatures are only necessary when control is weak and inexact.

Many diseases and some pests, are influenced to some degree by environment, but it is not possible to achieve control of most of them merely by environmental manipulation.

## Resistant strains of *Agaricus* spp.

There is no naturally occurring resistance to pests and few well-defined examples of resistance to diseases in strains of *A. bisporus*. It is possible that, among the many strains available to the commercial grower, some undiscovered differences in susceptibility do occur.

The best-known instances of resistance occur in *A. bitorquis* (page 98) where virus particles have not yet been recorded. At present, it is thought to be resistant to the viruses that affect *A. bisporus*. Conversely, a rapidly occurring soft rot caused by organisms apparently similar to *Pseudomonas tolaasii* has occasionally been observed on *A. bitorquis*, but not on *A. bisporus*.

Of the strains of *A. bisporus*, the cream (including off-white) and brown strains do not show such severe symptoms of virus diseases. This difference can be usefully exploited on farms where virus diseases have become endemic.

# 4

# Fungal Diseases

## Introduction

Fungi are the most important group of pathogens of the mushroom, occurring wherever the crop is grown. The incidence and severity of fungal pathogens varies from time to time: for instance, in the United Kingdom, *Mycogone perniciosa* was most commonly found in the 1950–60 period but, from the late 1960's to 1980, *Verticillium fungicola* var. *fungicola* predominated. At present, most fungal pathogens are fairly well controlled, although *Cladobotryum dendroides*, *Diehliomyces microsporus* and *Trichoderma harzianum* have become more common. Control is best achieved by careful farm management and, in particular, by attention to hygiene. Growers cannot rely on the genetic resistance of spawns nor on the continuing effectiveness of fungicides. Fungicides should be applied only to reduce inoculum levels so that routine hygiene can be successfully used to keep disease at a satisfactory level. Fungicides should never be used as a routine treatment as this may result in the development of fungicide-resistant strains of one or other of the pathogens. The fungicides in *Table 4.1* have label recommendations for use on the mushroom crop.

## Wet bubble or *Mycogone*

This disease, which is caused by the fungus *Mycogone perniciosa*, is common, although not often serious. When it develops early in a crop and is not controlled, it can cause considerable crop loss.

### SYMPTOMS

The most characteristic symptom of wet bubble is the development of distorted masses of mushroom tissue, which are initially white and fluffy but become

45

Table 4.1 Fungicides for mushroom diseases

| Common Names | Trade Name (and manufacturer) | Method and rate of Use | Fungi controlled |
|---|---|---|---|
| Benomyl* | Benlate (Du Pont) | Mix with casing or apply in place of first watering. Mix 240 g/100 m² (0.5 lb/1000 ft²) Drench 240 g/200 litres/100 m²(0.5 lb/ 40 gal/100 ft²) | *Dactylium, Mycogone, Trichoderma, Verticillium* |
| Carbendazim* | Bavistin wp, Bavistin DF/FL (BASF) | Mix with casing or apply in place of watering 250 g/250 ml/100m² of bed (0.5 lb/ 1000 ft²) when incorporating into casing. Apply 250 g/250 ml in 200 litres of water per 100m² of bed ½lb/8 fl oz in 40 gallons of water per 1000 ft²) when applied as the first watering. | *Dactylium, Mycogone, Trichoderma, Verticillium* |
| Chlorothalonil | Bravo 500 (SDS Biotech & BASF) Repulse (ICI) Sipcam Rover 500 (Sipkam) Tripart Faber (Tripart) BB Chlorothalonil (Brown Butlin) | Apply as a spray one week after casing and repeat if necessary not less than 2 weeks later Rate 220 ml in 100–200 litres/100 m² (7 fl oz/ 20–40 gal/100 yd²) | *Mycogone, Verticillium* |

| | | | |
|---|---|---|---|
| Prochloraz manganese | Sporgon (Darmycel) | Used as either a casing or spray treatment Use 120 g (4 oz)/100 m² (1000 ft²) of bed area to be cased or at first watering, in up to 180 litres (40 gallons). If required, apply again between first and second or second and third flushes; or, apply 60 g (2 oz) in 100 litres (40 gallons) 7 days after casing and after first or third flushes; or, 120 g/100 m² (4 oz/1000 ft²) in 180 litres (40 gallons) 7–9 after casing and between second and third flushes. | Dactylium, Mycogone, Verticillium |
| Thiabendazole* | Hymush (Agrichem Ltd) Tecto 60 WP (Merck, Sharp & Dohme) | Mix into casing or apply as a spray. Mix 180–250 g (6.5–8.5 oz) 200 litres (45 gal). When disease is serious, supplement above with 90 g (3 oz)/100 m²/200 litre between flushes or no initial treatment but weekly applications of 120 g (4 oz)/200 litres/100 m² NB: Maximum amount to be used is 750 g (24 oz) 100 m² to any one crop. A single initial dose should not exceed 250 g (8 oz) and supplementary doses 120 g (4 oz)100m²). | Dactylium, Mycogone, Verticillium |
| Zineb | Zineb 7% Dust (Hortag) | Occasional treatment after casing and between breaks Rate, 350 g/100m² (12 oz/1000 ft²) or, applied every week after casing throughout cropping and before watering 140 g/100 m² (4.5 oz/1000 ft²). | Dactylium, Mycogone, Red geotrichum, Verticillium |
| | Zineb Wettable (Hortag) | After casing and between flushes. 1 kg/1000 litres at rate of 5 litres/10m² (1 lb/100 gal at rate of 1 gal/100 ft²) | Dactylium, Mycogone |

* Benzimidazole fungicides

brown as they age and decay (*Figure 4.1*). These are known as 'sclerodermoid masses' and they may be up to 10 cm (4 in) across. They usually result from the infection of a sporophore initial and they take 10–14 days to develop. Small amber to dark brown drops of liquid develop on the surface of the undifferentiated tissue, especially in conditions of very high humidity *(Colour Plate 1)*. It is the wet decay and the shape of affected mushroom tissue that gives the disease its common name. In dry conditions the distorted masses remain dry in appearance and are very similar to those of dry bubble disease. In addition to the distortion symptom, small fluffy white patches of mycelium

**Figure 4.1** *Mycogone perniciosa* causing a large 'cauliflower-like' distortion. Such distorted mushrooms are referred to as sclerodermoid masses

may occur on the surface of the casing, following the infection of a developing mushroom below the casing surface.

When mature mushrooms are attacked, only the base of the stalk may be affected, causing a brown discoloration and, eventually, white fluffy mycelial growth. When affected, stem bases are often left in the bed after cutting the mushrooms and they are a source of inoculum. The stipe can sometimes become colonized, showing a brown streak, and the cap develops symptoms which appear as a sector of affected and stunted gills, covered by the characteristic white mycelium of the pathogen (*Figure 4.2*).

**Figure 4.2** A sporophore showing symptoms of partial colonization by *Mycogone perniciosa*. The pathogen has colonized one side of the stipe and radiated out on to the gills affecting a sector. The affected area shows the amber droplets of liquid typical of this disease

Infection at a very early stage in the development of the mushroom results in the sclerodermoid masses described above. The later the time of infection in the developmental stage of the mushroom, the less the distortion.

THE PATHOGEN AND DISEASE DEVELOPMENT

The pathogen produces predominantly two spore forms, one a single-celled, thin-walled and relatively short-lived conidium and the other a two-celled aleuriospore (terminal chlamydospore) with a very thick-walled terminal cell. It is believed that aleuriospores can survive for very long periods and there is circumstantial evidence suggesting that they remain viable for at least three years in stored casing. The conidia are light and may be airborne. The aleuriospores and conidia are spread by water splash. The fungus is thought to be fairly common in soils, and outbreaks of the disease often follow soil movement on the farm.

The primary source of the pathogen on most farms is contaminated casing. Generally, symptoms in the first flush indicate contamination of the casing. Compost is not an important source. Spores may survive on the surfaces of buildings, or may be carried in crop debris and, in this way, can contaminate crops. There is no evidence that *Mycogone* spores are commonly spread by flies or other insects. Once the pathogen is established in the crop, the main means of spread is by water splash and by excess water running off the beds. Pickers may also spread the pathogen on their hands, on tools, boxes and clothing. There is no evidence that this pathogen will grow in the casing or compost and diseased mushrooms only result when initials develop adjacent to spores. Contamination of mushroom compost, or even the lower layers of the casing, is not, therefore, significant.

CONTROL

All commercially available strains of *A. bisporus* are susceptible to wet bubble. There have been very few reports of *A. bitorquis* being affected but this may reflect the fact that this species is grown less often. One of the most important means of control is the elimination of primary sources of the pathogen. This can be achieved only by paying strict attention to hygiene (*see* Chapter 3). It is particularly important to ensure that the casing materials are stored in an area where they will not become contaminated by debris or dust from cropping houses (*see Figure 3.3*).

Once the pathogen has become established in a crop, spread must be minimised. All affected mushrooms should be carefully removed by a special team of pickers. The chance of finding and removing all diseased mushrooms is greatly increased if the beds are well illuminated. Pickers should wear gloves and use a disinfectant regularly, putting the diseased mushrooms into a bucket of dilute disinfectant. All diseased mushrooms should be removed before the crop is picked; covering affected mushrooms with salt or pots is an alternative to removing them.

As watering is one of the most important means of spread, this should be done only after all the diseased mushrooms have been removed. If a special team of workers is employed to remove diseased mushrooms, it is often convenient to water immediately after this operation; alternatively, watering can be left until all the healthy mushrooms have been harvested.

If plastic pots or salt are used to minimise spread, it is essential to push the pots well into the casing, preferably down as far as the compost surface, otherwise sideways drainage of water will disperse the spores of the pathogen. The salt is used to cover the affected mushrooms and surrounding area of casing.

Prochloraz manganese gives a good control of wet bubble, as do the benzimadazole fungicides (benomyl, carbendazim, thiabendazole); there are no records of resistance to any of these fungicides in the United Kingdom. But disease control with benomyl has failed on some farms and this has been associated with the breakdown of the fungicide in the casing, rather than with fungicide resistance. In such cases, it is worth using thiabendazole, which persists in the casing longer, or prochloraz manganese. Dithiocarbamates, such as zineb and maneb, and also chlorothalonil, will give some control, especially when inoculum levels are low. Details of the fungicides available are given in *Table 4.1* (page 46).

If the casing is contaminated, *Mycogone* can be eliminated by treating the casing before use with 1% formalin (using 25 litre/m$^2$). Treatment of the casing with formalin can seriously affect cropping if done carelessly and, for this reason, it is often considered only as a last resort. A safer treatment, but one which is more expensive to apply, is the treatment of the casing with heat, raising the temperature to a maximum of 82°C (180°F) for about 30 minutes. Steam air mixtures and experimental micro or radio wave treatment have been used for this purpose.

ACTION POINTS

- Observe strict hygiene at all times (*see* Chapter 3);
- Make sure the casing is mixed and stored in a clean area;
- When risk of *Mycogone* is high, use benomyl or prochloraz manganese applied to the casing;
- Where benomyl gives inadequate control, use thiabendazole or prochloraz manganese;
- Where control of *Verticillium* is also a consideration, prochloraz manganese should be used.

## Dry bubble or *Verticillium*

This is probably the most common and serious fungal disease of the mushroom crop. If left uncontrolled, dry bubble can totally destroy a crop in 2–3 weeks. In the United Kingdom, it is caused by the fungus *Verticillium fungicola* var. *fungicola* (syn. *V. malthousei*), although a closely related fungus, *V. psalliotae*, causes some similar symptoms. Recent work has shown that there are other varieties of *V. fungicola*. A second form, *V. fungicola* var. *aleophilum*, may be more commonly associated with crops of *A. bitorquis*, or may occur on *A. bisporus* in countries where ambient temperatures are high.

SYMPTOMS

The symptoms caused by *V. fungicola* var. *fungicola* are varied and depend upon the development stage of the mushroom at the time of infection. When mushrooms are infected at an early stage in their development, they develop into small undifferentiated masses of tissue up to 2 cm (1 inch) in diameter (*Figure 4.3*). When affected at a later stage, the mushrooms are often imperfectly formed with partially differentiated caps, or with distorted stipes and

**Figure 4.3** Small, undifferentiated masses of mushroom tissue following infection by *Verticillium fungicola* var. *fungicola*. Some affected sporophores develop small pimples on the cap

with tilted caps (*Figures 4.4a, b, Colour Plates 2 and 3*). Such affected mushrooms are covered in a fine grey-white mycelial growth and, although discoloured, are dry and do not rot. Occasionally, more fully differentiated mushrooms are affected and these show small, pimple-like outgrowths from

**Figure 4.4a** An imperfectly formed mushroom resulting from attack by *Verticillium fungicola* var. *fungicola*.

**Figure 4.4b** Peeling of the stipe is a characteristic symptom of dry bubble disease (*Verticillium fungicola* var. *fungicola*)

the top of the cap, or blue-grey spots (1–2 cm in diameter) on the cap surface (*Figure 4.5*). Such spots often have a yellow or bluish-grey halo around them.

*V. fungicola* var. *aleophilum* does not generally produce distortion, but does affect the cap, on which it produces a dark brown spot (*Figure 4.6, Colour Plate 4*). These spots are similar to those of bacterial blotch but differ in the appearance of the edges of the lesions, having an indistinct margin and being generally darker in colour. The dark brown spots may eventually become covered by a white 'bloom' as the pathogen produces spores. *V. psalliotae*, like *V. fungicola* var. *aleophilum*, does not produce distorted mushrooms but does give cap-spotting symptoms very similar to that caused by *V. fungicola* var. *aleophilum*.

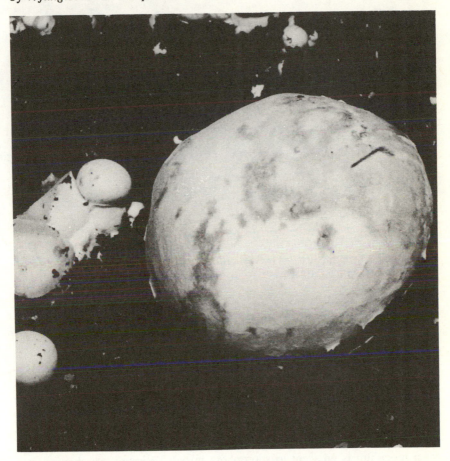

**Figure 4.5** Diffuse spotting of the mushrooms caused by *Verticillium fungicola* var. *fungicola*. The spots vary in colour from pale brown to blue-grey

**Figure 4.6** Cap spotting caused by *Verticillium fungicola* var. *aleophilum*. The spots are dark brown with a diffuse edge

THE PATHOGEN AND DISEASE DEVELOPMENT

All *Verticillium* species produce thin-walled conidia. There are no other known spore forms. It is also possible that resting mycelium may enable the pathogen to survive in a dormant state for considerable periods. Contaminated casing, but not compost, is probably the most common initial source of *Verticillium* on mushroom farms. Like *Mycogone*, there is no clear evidence that this pathogen grows through the casing or compost. Disease develops when inoculum is adjacent to sporophore initials. Primary introduction may also be by airborne spores and also by spores carried by flies, mites or pickers. Unlike the spores of *M. perniciosa*, conidia of *Verticillium* are produced in clusters surrounded by sticky mucilage, and it is this mucilage which enables them to become attached to dust, flies, mites, debris and pickers. Once these conidia have become attached to the hands, it is almost impossible to remove all of them by the ordinary process of washing, even using hot soapy water; distribution on hands and clothes is one of the most important means of transfer, both within and between crops. Some mites (*Tyrophagus* spp.) are known to feed on the spores and mycelium of *V. fungicola* var. *fungicola*, and viable spores have been recovered from faecal pellets. Such infested mites could be carried from house to house, or from farm to farm. Watering is also an important means of dispersal, as excess water running off the beds carries the spores to lower beds, or to the floor of the house. When the floor dries, air movement over its surface may cause the spores to become airborne.

Although dry bubble occurs over a wide range of environmental conditions, the optimum temperature for disease development is about 20°C (68°F). The period from infection to symptom expression at 20°C is about 10 days for the distortion symptoms and 3–4 days for cap spotting. The pathogen grows best at 24°C (75°F). In contrast, *V. fungicola* var. *aleophilum* and *V. psalliotae* grow best at higher temperatures (27°C, 81°F). This may account for the infrequency of occurrence of the latter two pathogens in crops of *A. bisporus* in the United Kingdom and their greater frequency in crops of *A. bitorquis*, which are grown at higher temperatures than *A. bisporus*.

CONTROL

None of the commercially available spawns are resistant to dry bubble.

Disease appearing in the first flush indicates effective airborne spread by flies or mites or spread by workers, or it can indicate contaminated casing. If it is thought that the casing is the source, a careful examination of the storage conditions of the ingredients is essential, as is thorough disinfection of all possible surfaces. The casing can be heat treated to eliminate spores of the pathogen (*see Mycogone* control).

Strict attention to hygiene is of paramount importance at all times for effective control of dry bubble. Because of the easy distribution of the spores by pickers, mites, flies, and probably mice and rats, it is vitally important to remove all affected mushrooms at the earliest possible stage. Affected pinheads are easily missed, especially if the lighting of the bed surface is inadequate. A team of pickers whose sole job is to remove diseased mushrooms should inspect the beds daily, carefully removing all affected mushrooms and placing them in buckets of disinfectant. Rubber gloves should be worn so that hands can be dipped in the disinfectant at regular intervals. Spread can also be prevented by covering affected sporophores with salt or pots. It is essential to push the pot well into the casing as far as the surface of the compost in order to prevent the sideways drainage of water, which would distribute spores. When common salt is used to minimize spread, it is heaped over the affected mushrooms, killing both the affected mushroom and the pathogen. Flies and mites must be controlled, particularly in the summer months when fly infestations are likely to build up quickly.

If levels of disease become excessively high, it is advisable to terminate the affected crop either by cooking-out or drenching with a disinfectant, such as formalin, or a phenolic material. It is particularly important to ensure that the cook-out or chemical treatment at crop termination is effective. A temperature of 70°C, maintained for ten hours, should ensure complete kill of all the spores on surfaces. Great care must be taken to avoid the dispersal of debris when transporting spent compost. It is essential to take all spent compost well away from the farm.

Prochloraz manganese gives a good control of dry bubble disease and there are no records of fungicide resistance. This fungicide should not be used routinely as this will maximise the risk of resistant strains arising. Almost all populations of *V. fungicola* var. *fungicola* are resistant to the benzimidazole fungicides; strains resistant to one of the benzimidazole fungicides are also resistant to the others. Chlorothalonil and zineb are also alternative fungicides, but these are moderately effective only when the disease levels are low (*see Table 4.1*).

ACTION POINTS

- Observe strict hygiene throughout the farm;
- Make sure the casing ingredients are stored and mixed in a clean area;
- Control flies and mites;
- Remove all affected mushrooms before picking and watering;
- When disease risk is high, apply prochloraz manganese;
- Terminate severely affected crops early by either cooking-out or the use of effective disinfectants.

## Cobweb or *Dactylium*

Although not uncommon, this disease is rarely epidemic, but can cause losses of crop and is often difficult to eliminate totally. It is caused by the fungus *Hypomyces rosellus* (stat. conid. *Cladobotryum dendroides* syn. *Dactylium dendroides*). Another species, *H. auranteus* (*Cladobotryum variospernium*) also causes disease in mushroom crops.

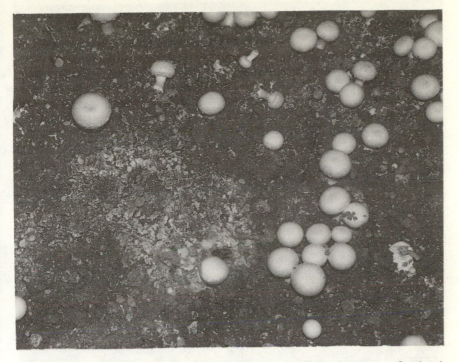

**Figure 4.7** A patch of mycelium of *Hypomyces rosellus*, the cause of cobweb disease. The surrounding mushrooms are not affected

SYMPTOMS

The disease caused by *H. rosellus* is characterized by the growth of coarse mycelium covering affected mushrooms, hence the common name of cobweb disease. Mushrooms can be attacked at any stage in their development, as the pathogen will rapidly colonize the mushroom and surrounding casing. Roughly circular patches of dense white mycelium may appear on the casing surface, with no other obvious sign of the disease (*Figure 4.7*). Affected mushrooms eventually turn brown and rot (*Figure 4.8a*). In time, the mycelium of the pathogen may change colour, becoming pink or red, and the cobweb appearance is replaced by a mat of mycelium (*Figure 4.8b, Colour Plate 5*). Recently, brown or pink-brown spots on the sporophores, with a poorly defined edge, have been associated with cobweb disease (*Figure 4.9, Colour Plate 6*). This symptom has occurred commonly where prochloraz manganese has been used. *H. auranteus* produces similar symptoms but the mycelium of the pathogen does not develop the red colour seen with *H. rosellus*.

**Figure 4.8a** A mushroom becoming colonized by *Hypomyces rosellus*. The cobweb-like mycelium of the pathogen has already covered the gills

THE PATHOGEN AND DISEASE DEVELOPMENT

Conidia are the only commonly occurring spore form of this pathogen. The spores are relatively large, multicellular and produced on verticillate conidiophores. They are dispersed mainly by water splash, in drainage water and, less commonly, by pickers; there is no evidence that they are distributed by flies. Spread of the pathogen can also occur by growth over the casing surface. Fragments of mycelium from the casing surface may also be distributed by pickers.

*Hypomyces* spp. are soil-inhabiting fungi and may be introduced into the casing with soil, or by spores or mycelium spread on debris, or by workers. Unless the spore inoculum level is high, it is usual for the first symptoms of the disease to occur in the later flushes of the crop. Even when the casing is severely contaminated with spores, disease symptoms may not develop until

**Figure 4.8b** Extensive colonization of mushroom sporophores by *Hypomyces rosellus* producing the most characteristic cobweb symptoms. The white mycelium of the pathogen eventually changes colour and becomes pink or red

the fourth or fifth flushes. However, contamination of casing with cobweb mycelium results in a rapid development of symptoms, which can then appear in the first flush. *H. rosellus*, unlike *Verticillium* and *Mycogone*, is capable of growth on or through the casing. It is possible that more than one species of *Hypomyces* commonly occurs in the United Kingdom, accounting for the variability seen in the symptoms of the disease on different farms.

CONTROL

None of the commercial spawns are resistant to cobweb. The disease is easily controlled by hygiene and by the use of fungicides: as soon as it occurs, the affected mushrooms should be carefully removed, together with an area of casing from around the affected sporophores. Alternatively, the whole of the affected area, and some clean casing around it, can be covered with common salt.

**Figure 4.9** Pale brown or pink-brown cap spotting can be caused by *Hypomyces rosellus*. This symptom has only recently been recognized as part of the syndrome of cobweb disease

There are no records of fungicide resistance to any of the commonly used fungicides. Cobweb disease is well controlled by the benzimidazole fungicides, which are best used as spot treatments at an early stage in disease development, or as an overall spray for more extensive attacks (*see Table 4.1*). If the casing is suspected as being the source of the pathogen, a benzimidazole spray applied before the first flush will give effective control. Benomyl degradation may make it ineffective, and thiabendazole is a worthwhile alternative.

- Remove affected areas of casing as soon as they appear and spot treat with a benzimidazole fungicide (*see Table 4.1*);
- Avoid the transfer of contaminated casing to healthy crops by strictly regulating the movement of pickers;
- When cap spotting occurs, examine the spray programme and change to a benzimidazole material; one application, immediately after cap spotting is seen, should suffice.

## False truffle or truffle

False truffle (*Diehliomyces microsporus* syn. *Pseudobalsamia microspora*) is one of the most troublesome and persistent of the competitor moulds. This fungus not only competes with the mushroom mycelium for food and space but is believed to attack mushroom mycelium and cause mycelial death.

SYMPTOMS, THE PATHOGEN AND DISEASE DEVELOPMENT

The first symptom of false truffle is often the patchy appearance of the crop, with areas of the affected bed failing to produce mushrooms. The mycelium of *Diehliomyces* is white, becoming cream in colour with age, and often grows in dense wefts. Initially, it is not always easily distinguishable from the mycelium of the mushroom, but circular areas of white, coarse, fluffy mycelium on the compost surface during a spawn run are characteristic. The ascocarps of the fungus (*see below*) form within the dense wefts within 15–21 days, depending on the temperature, and are often the only indication of the presence of the mould as they persist after the weft of mycelium has disappeared. Where the *Diehliomyces* mycelium has grown, the mushroom mycelium disappears, and the compost is black, often wet, and is said to have a characteristic chlorine-like smell. The ascocarps vary in length from 3 mm to 40 mm (0.1 to 1.5 in) and are pale yellow to cream, turning red-brown as they age. The exterior corrugation on their surface gives them a characteristic appearance, often said to resemble a shelled walnut or the surface of a brain (hence the once-common name 'calves' brains'; *Figure 4.10*). The ascocarps are frequently most numerous at the casing-compost junction, along the sides of the boxes or beds, or against the base of the bed but can also occur anywhere within the casing or compost, including on the surface of the casing. They also form on the outsides of spawned blocks or bags of compost next to the plastic cover. The ascocarps contain many small ascospores about 5 μm in diameter and, following the breakdown of the ascocarps by bacteria, the ascospores are released into the casing or compost. Spread of the ascospores is most likely to occur in drainage water and on airborne debris. *D. microsporus* is said to be a

**Figure 4.10** The ascocarps of *Diehliomyces microsporus* which are found within the compost and often at the sides of boxes or blocks.

common soil dweller and severe outbreaks on farms often follow operations which involve soil movement.

The optimum temperature for the germination of *Diehliomyces* ascospores is 30°C (86°F), and is stimulated by the presence of actively growing mushroom mycelium. The spores will, however, germinate at much lower temperatures (16°C, 61°F). Until recently, the ascospores were thought to be very heat and chemical tolerant, but some recent studies have indicated that they do not have exceptional properties, and are killed by a 30 minute treatment at 60°C (140°F). It is likely, however, that they can withstand very high temperatures if they are dry when treated. This may account for other reports which suggest that treatment at 80°C (176°F) for 120 minutes is not enough to kill them.

For severe crop loss to occur, *Diehliomyces* must be present in the compost at, or during, spawn-running, and then yield reductions can be as high as 75%. High compost temperature at this stage, e.g. 27–28°C (80–82°F), greatly favours truffle development. A short period at this temperature, followed by a return to lower temperatures appears to act as a stimulus for the development of the fungus. In experiments, contaminated compost kept at 15°C (59°F) was not affected by false truffle. Recently, crops of *A. bitorquis* have been seriously affected by false truffle: the higher growing temperature used for this species may make it more prone.

CONTROL

All *A. bisporus* spawns are susceptible. Some strains of *A. bitorquis* (K26 and K32) are reported to have some resistance.

Strict attention to hygiene, an effective peak-heat and installation of filters (2 µm diameter mesh) in the peak-heat and spawn-running areas should prevent false truffle occurring (*see* Chapter 3). High compost temperatures during spawn-running must be avoided as far as possible as these will encourage germination of ascospores. Covering the spawn-run with paper which is regularly sprayed with formalin will minimize the risk of contamination at this stage (*see* page 96). If false truffle is a persistent problem on a farm, control can be achieved by keeping the compost temperature as low as possible, especially during the first week of spawn-running. This is often difficult to achieve and also markedly delays the speed of development of the crop. So, although feasible, it is not a very practical way of controlling false truffle.

As soil is a potential source of the pathogen, it is important to ensure that compost is not contaminated. The composting area must have a smooth concrete surface to allow thorough disinfection between batches of compost. All affected crops must be properly cooked-out. A temperature of 70°C (158°F) maintained for 12 hours, or 80°C (176°F) for 2 hours, is believed to be adequate to kill ascospores, but it is important to make sure that this temperature is achieved throughout the compost. In addition, wooden boxes and wood used to make beds must be thoroughly cleaned with a disinfectant at the end of cropping. Metal shelving can be thoroughly disinfected with formalin after emptying, and thorough net cleaning is also essential.

There are no satisfactory fungicidal treatments of compost or casing, either before or after the disease has occurred.

ACTION POINTS

- Prevent soil contamination of compost by ensuring that the compost yard is adequately disinfected between batches of compost;
- Use absolute filters during peak-heat and spawn-running;

- Avoid high temperatures during spawn-running;
- Ensure that peak-heat and cook-out temperatures are evenly maintained;
- Remove all debris from boxes and chemically treat between every crop.

### *Trichoderma* diseases

*Trichoderma* is characterized by the production of very large quantities of dark green spores and it is these that are seen during various stages of cropping. A number of different species and strains of *Trichoderma* are found in mushroom culture, some harmless and others very damaging. The relationship between the various *Trichoderma* spp. and the mushroom is not fully understood and almost certainly varies with the species and strain, but some are known to be pathogenic. The genus is well known for its mycoparasitic ability and for the production of toxins and antibiotics. The species most frequently associated with the mushroom crop are *Trichoderma aureoviride, T. harzianum, T. koningii, T. pseudokoningii* and *T. viride*.

SYMPTOMS

The symptoms vary with the species. Green mould frequently occurs on the woodwork of shelves and boxes, particularly after they have been heat treated. New wood is particularly likely to be affected. The mould growth begins as a prolific development of white mycelium which, within 2–4 days, turns green as the spores are produced. All the above species have been found, but *T. viride, T. koningii* and *T. pseudokoningii* are probably more commonly found than *T. harzianum*. Extensive colonization of the wood of trays often leads to cap spotting of the mushrooms, particularly those developing near the edges. These spots are pale brown in colour without a clearly defined edge. They are often numerous but generally less than 5 mm in diameter. It is not known whether all of the above species cause spotting, but *T. koningii* and *T. pseudokoningii* have both been associated with severe outbreaks in the United Kingdom *(Colour Plate 7)*.

A similar range of species has been found on the cut ends of mushrooms. Growth of this form of green mould is common but rarely appears to be associated with a disease problem. The most damaging form of green mould is that associated with colonization of the compost at the same time as the mushroom mycelium. A dense white mass of mycelium can often be seen at an early stage in crop development, either through the base boards of boxes, or through the netting on shelves. This mould also appears on the casing surface and, generally by the first flush stage of the crop, it turns dark green as spores are formed *(Colour Plate 8)*. This fungus is generally a strain of *T. harzianum*, which is characterized by its fast growth rate (more than 1 mm per hour at 27°C (81°F) on agar), by a longer vegetative phase than other isolates of this species, and a higher optimum temperature for growth (27°C). Where it colonizes mushroom compost, mushroom yields are reduced. Associated with outbreaks of this disease are very large populations of red pepper mites (*see* page 12), which feed on the mycelium and spores of the *Trichoderma*. Sometimes the red pepper mites are the primary indication of the presence of *Trichoderma*. Other strains of *T. harzianum* are also found in mushroom culture and may grow in the compost to some extent but appear to have less effect on yield or quality.

THE PATHOGEN AND DISEASE DEVELOPMENT

*Trichoderma* species are commonly found in soil and on organic material in many different habitats. They produce chains of conidia, which are very readily dispersed. The perfect stage of the fungus, *Hypocrea*, is less commonly seen. Spores can be readily spread by air movement, by insects and mites (particularly red pepper mites in the case of *T. harzianum*), by personnel and on all the various containers and equipment used on a mushroom farm. The growth requirements of various species are not the same, and the optimum growth temperature varies from 22°–26°C. Some *Trichoderma* spp. grow particularly well at pHs below 6, especially if the nitrogen level is low. A carbon:nitrogen ratio of 22 or 23:1 appears to favour the growth of *Trichoderma* in compost (the normal ratio is 15:1). Acidifying normal phase II compost with acetic acid or adding sugar can turn it into a very suitable medium for the growth of *Trichoderma*.

The *T. harzianum* strain which has been shown to be associated with severe crop loss has been designated strain 2 (Th2). This fungus grows particularly well in the presence of mushroom mycelium and thus causes the greatest problem when it is introduced into the compost at the same time as the spawn. It has become particularly serious in enclosed systems of spawn running, for instance, in bags or in wrapped blocks where temperatures are frequently high. Rats and mice feeding on surface spawn may introduce and spread the pathogen. Populations of Th2 in an affected compost are generally at least 1 million propagules per gramme. Fully run compost inoculated with Th2 does not become affected. Light is required for the production of spores and Th2 appears to require more light than other strains, although this is probably not significant in mushroom production.

CONTROL

The control of *Trichoderma* on wood and on the surface of casing is generally not necessary, except when it is associated with cap spotting. Spraying the woodwork during the end of the spawn-running period with benomyl will usually reduce the growth of the fungus. Benomyl or carbendazim applied to the crop — as for the control of wet bubble disease — will also control cap spotting. The routine application of this fungicide should not be encouraged because of the likelihood of fungicide resistance developing.

Control of Th2 is much more difficult. Spore contamination is widespread on affected farms and distribution is made very effective by the movement of personnel, insects, mites and small animals, such as mice and rats. The aim of the control measures is to eliminate the inoculum. This can only be done by the strictest attention to hygiene and a programme, similar to that adopted for mushroom virus disease control (*see* pages 94–98 and *Table 3.3*), is necessary. Particular attention should be given to the prevention of contamination of compost from cool-down to the first week of spawn-running. Th2 spores are killed by normal peak heat temperatures but air used after the kill temperature has been reached must be filtered. The spores are not smaller than 4 μm in diameter, and are therefore removed by absolute filters. Trays must be thoroughly cleaned before use, either by cooking-out at the end of the crop, or by combining this with chemical treatment. Treated trays readily become recontaminated if left standing around on the farm between use. Spawning lines, conveyors and transportation equipment must be thoroughly surface sterilized with 2% formalin immediately before use. It is crucial that the spawn is handled cleanly and that the spawn-hopper is surface sterilized with formalin before use. Regular disinfection of all structures and concrete surfaces around the farm is essential. Patches of red pepper mites should be covered in salt or burnt-off with a flame gun as soon as they appear in order to prevent the mites carrying spores into spawn-runs. Pickers' overalls should be cleaned daily, either by machine washing or by heating them in a tumble drier for 30 minutes on maximum setting.

ACTION POINTS

- Filter air into phase II rooms and spawn-running rooms;
- Use 2% formalin on the spawning-line before use;
- Disinfect the spawn-hoppers and make sure that spawn is handled with clean hands;
- Control rats and mice;
- Cook-out empty boxes at 70°C for 1 hour and/or dip in disinfectant;
- Salt-out patches of red pepper mites;
- Cover spawn-runs with paper sprayed every 3 or 4 days with 2% formalin.

## Shaggy stipe

This disease, which is uncommon and causes very little loss in yield, is caused by the fungus *Mortierella bainieri*.

SYMPTOMS

The most characteristic symptom of shaggy stipe is the peeling of the stalk of the affected mushrooms, giving a shaggy appearance (*Figure 4.11*). The stalk and cap are usually discoloured, becoming dark brown as the disease progresses. The coarse grey-white mycelium of the pathogen can usually be seen growing over the affected mushroom tissue and also over the surrounding casing. It is superficially similar to cobweb (*see* page 59), but is distinguishable from it by the colour of the mycelium and the symptoms shown by the affected mushrooms. The cap may also develop a brown blotch, often surrounded by a yellow ring.

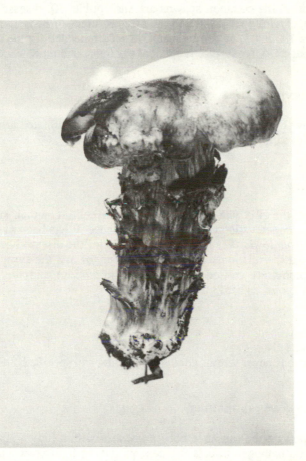

**Figure 4.11** Shaggy stipe caused by *Mortierella bainieri*. Affected mushrooms eventually become sparsely covered by coarse grey mycelium of the pathogen

*M. bainieri* produces masses of sporangia and sporangiospores and it is these that spread the pathogen. It is likely that the spores are both airborne and waterborne. The fungus is a common soil dweller and soil is likely to be the main initial source. The most severe outbreaks of this disease occurred during the period when benzimidazole fungicides (*see Table 4.1*) were used extensively.

CONTROL

There is no information on the resistance of spawns to this pathogen, but it is likely that all the common spawns are susceptible. All diseased mushrooms should be carefully removed and the affected areas of beds spot-treated with zineb (*see Table 4.1*) or common salt, as recommended for the control of dry bubble (page 58). No further fungicidal treatment should be necessary.

ACTION POINT

● Remove diseased sporophores and spot-treat the affected area of bed with zineb or salt.

## Gill mildew

Stunting of the gills with a growth of white mycelium on their surfaces is said to be the characteristic symptom of this disease. It has been suggested that a fungus (*Cephalosporium* spp.) is responsible for the disease but there is some doubt about the relationship between this fungus and the symptoms. Similar gill symptoms are caused by *Mycogone perniciosa* and the physiological condition known as hard gill (*see* page 157).

CONTROL

This disease is never serious enough to warrant any specific control measures.

### *Aphanocladium* cap spotting

Brown spots on the cap have been associated with a fungus *Aphanocladium album*. The pathogen affects the cap, producing light-brown to dark-brown roughly circular spots, which may be up to 10 mm in diameter (*Colour Plate 9*). Under conditions of high relative humidity, white aerial mycelium is formed on the affected tissues. Contamination of the casing can result in 40–50% of mushrooms being affected. Very similar symptoms are caused by *Verticillium* spp. (*see Figure 4.6*).

CONTROL

This disease is occasionally serious. Fungicides recommended for the control of dry bubble disease (page 58) are likely to give control (*see Table 4.1*).

## *Hormiactis* cap spot

There have been a number of reports of the fungus *Hormiactis alba* causing spotting in mushroom crops. The symptoms consist of irregular brown spots or blotches about 10 mm in diameter occurring anywhere on the cap surface. This disease has never been serious. *Hormiactis* cap spot is probably the same disease as that previously attributed to a species of *Ramularia*. It is readily controlled by spot treatment with salt or a benzimidazole fungicide (*see Table 4.1*).

# 5

# Bacterial Diseases

## Bacterial blotch or brown blotch

Blotch is one of the most common and serious diseases of *A. bisporus* and is responsible for considerable losses of mushrooms every year. The disease also affects *A. bitorquis*. It is caused by a bacterium (*Pseudomonas tolaasii*) which is very widespread on mushroom farms. When conditions are favourable, epidemic levels of blotch occur. Crops grown in the autumn months are often the most severely affected.

SYMPTOMS

The most characteristic symptom of bacterial blotch is the occurrence of brown areas or blotches on the surface of the cap. These may be initially light in colour (*Figure 5.1a*) but may eventually become dark brown (*Figure 5.1b*). Severely affected mushrooms may be distorted and the caps may split where the blotch symptoms occur (*Figure 5.1a, Colour Plate 10*). The stalk may also be affected. The distribution of symptoms on the cap often coincides with the parts of the cap which remain wet the longest, i.e. the edges or parts where two mushrooms are touching.

DISEASE DEVELOPMENT

On many farms the pathogen appears to be endemic, probably surviving between crops on surfaces, in debris, on tools and on various structures. It is also a natural inhabitant of both the peat and chalk. When the disease is present, bacteria are readily moved from crop to crop on the hands of pickers and on boxes, ladders and other implements. Sciarids and mites are also important in spreading the pathogen. Thus, the bacteria can be dispersed on debris, on mushroom spores, in water droplets, by flies and on the hands of

**Figure 5.1a** Small light brown spots on the cap surface, typical of bacterial blotch (*Pseudomonas tolaasii*)

**Figure 5.1b** Dark brown spots of bacterial blotch

pickers; in addition, once the disease has become established in a crop, watering will disperse the bacteria very readily.

Mushrooms often become infected at a very early stage in their development. Infection depends on a high population of the bacterium and, once the critical level is reached, infection occurs. The enlargement of the microscopic spots on the cap which follow infection is dependent upon environmental conditions, and is favoured by temperatures of at least 20°C (68°F), together with the presence of water. Fluctuating temperatures at high relative humidities can result in condensation on the cap surface, resulting in long periods of cap wetness. Such conditions frequently occur in the autumn, and are very favourable for blotch development.

A high proportion of mushrooms can be affected once the pathogen is established in the crop and the bacterial population in the casing is above the critical level. The extent of the damage can be minimized by adjusting the environment. However, if infected but symptomless mushrooms are cut and stored in conditions favourable for lesion development, spots will appear and enlarge. Thus, mushrooms that are apparently healthy at harvest can show marked symptoms a few days later.

**Figure 5.1c** Cap splitting resulting from a severe attack of bacterial blotch

CONTROL

None of the commercial strains are resistant to the disease. The casing is most likely to be one of the primary sources of the pathogen, so it is important to ensure that the ingredients are 'clean'. Where the disease is a persistent problem, treating the casing before use with formalin or heat may be beneficial.

As high relative humidity and surface wetness encourage symptom expression, the environmental conditions should be adjusted to avoid such conditions. A Piché evaporimeter has been found by some growers to give a useful guide to the rate of evaporation within the crop. With this simple apparatus, it has been shown that blotch is rarely a problem if the daily rate of evaporation is at least 0.8 ml of water (*Figure 5.2*).

Chlorine was used as a routine spray to the beds to help to keep the bacterial population down to such a level that blotch did not result. To be successful, such a treatment must start at casing, and not to wait until disease symptoms appear. Various preparations of sodium hypochlorite have been commonly used, and these usually contain about 10% available chlorine. As the chlorine content decreases once the container has been opened, it is important to buy quantities that can be used quickly. If preparations of

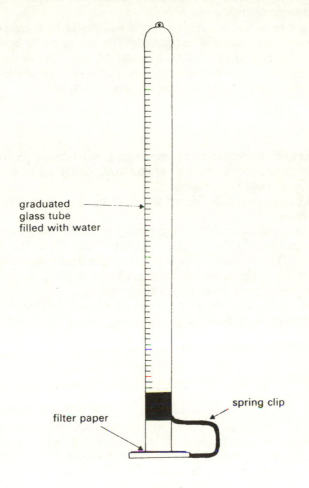

graduated
glass tube
filled with water

spring clip

filter paper

Piché evaporimeter

**Figure 5.2** A Piché evaporimeter. This simple apparatus gives a very useful guide to the rate of evaporation within a crop which can be used to identify conditions conducive to the development of bacterial blotch (apparatus supplied by Cassella London Ltd., Regent House, Britannia Walk, London N1 7ND)

hypochlorite have to be stored, they are best kept in a refrigerator or cold store and away from the light. The available chlorine content of a preparation can be checked if there is any doubt (*see* page 33). The chlorine is used at 150 ppm applied at every watering (i.e. 150 ml of hypochlorite with 10% available chlorine added to 100 litres of water, or 24 fluid oz of hypochlorite per 100 gallons of water). New pesticide regulations do not permit, at present, the use of chlorine on mushroom beds.

There have been various reports of the successful biological control of bacterial blotch using bacterial antagonists. The antagonist is applied to the casing from the beginning of the crop and prevents the build-up of the pathogen population. No preparations are commercially available in the United Kingdom, but the product Concord is sold in Australia.

ACTION POINTS

● Avoid surface condensation on developing mushrooms as, under such conditions, blotch will develop, almost irrespective of how thoroughly other aspects of control are applied;
● Store all casing materials, before and after mixing, in an area free from contamination;
● Adjust the conditions within the cropping house so that, whenever possible, evaporation is taking place from the surface of the developing mushrooms. This is particularly important immediately after watering;
● Make sure that temperature control is as precise as possible, as stable temperatures will prevent the dewpoint being reached;
● When the disease is established, remove all affected mushrooms and apply measures to prevent the pathogen spreading on pickers' hands and/or by watering.

## Ginger blotch

This disease, caused by *Pseudomonas gingeri*, which is closely related to the organism which causes bacterial (brown) blotch (*P. tolaasii*), has only recently been described. The ginger colour of the blotches, which do not change with age, distinguishes this disease from brown blotch. As far as is known, the ecology of the organism is similar to that of *P. tolaasii*, and the casing materials are probably the most important primary source. Control measures are identical to those recommended for bacterial blotch.

**Figure 5.3** Mushrooms showing some of the typical symptoms of mummy disease, in particular the elongation of the stipes, the small caps and the cap tilting (photograph kindly supplied by Dr Simon Oxley)

## Mummy disease

Mummy disease is not common, although it can cause large crop losses. Some farms have a persistent problem and, in such cases, losses over a series of crops can be considerable.

There is still considerable doubt about the cause of this disease. Although both American and Dutch work has indicated that a bacterium (a pseudomonad possibly related to *P. tolaasii*) may be responsible, experiments to prove the pathogenicity of isolates have either failed, or been only partly successful. Because of these problems, it is difficult to identify mummy disease positively.

SYMPTOMS

The symptoms of mummy disease are fairly distinctive, although there is some overlap with those attributed to virus diseases (*see* pages 84–85) and an abiotic disorder (page 163). The most characteristic feature of the disease is its fast rate of spread, which is quoted as being 10–25 cm (4–10 inches) bed length per day. When crops are grown in trays, bags or blocks, spread is less obvious, unless there is mycelial contact between trays. Affected mushrooms die and become very dry, with the internal tissue discoloured, often with brown streaks. When cut across, the affected stipes sometimes show small pinhead-sized dark brown spots. The cap may sometimes be distorted and is commonly tilted (*Figure 5.3*). At the base of the stalk the mycelium is very stringy, and there is often a basal swelling, together with a growth of fluffy white mycelium. When affected mushrooms are removed from the bed, a large amount of casing adheres to the base of the stipe. It is said that pickers are able to detect affected mushrooms by the feel of the stipes when they are cut or trimmed, because of the dryness of the tissue.

As the crop is frequently grown in discrete units, the rate of spread — which is so characteristic in a bed system — is not apparent. However, one other feature of mummy disease-affected crops is the complete failure of subsequent flushes. In this respect, mummy disease is very similar to severe virus disease.

The only way in which the disease has been reproduced consistently is by placing compost or casing from an affected bed into clean compost at the time of spawn-running; there is no evidence of spread in any other way.

Although no comparative studies have been made, the disease has been said to occur more commonly in some spawn strains than others. Whether this is because it is favoured by the growing conditions specifically used for such strains is not known. There are some indications that mummy disease is most severe when the compost is wet, or becomes excessively so, during peak-heat.

CONTROL

As soon as mummy disease is identified, it may be possible to localize it in shelves by digging a trench across on either side of the affected area and about 1.5 m (c. 1.5 yd) in advance of the symptoms. The trench should be 20 cm

(8 in) wide, and all the compost should be thoroughly removed or placed on the surface of the affected area. The isolated area of bed and the wood exposed in the trench should be drenched with 0.5% formalin, and the surface of the affected area should be covered with polythene. With a tray crop, care should be taken to isolate the affected trays, making sure that there is no mycelial contact between them and adjacent healthy trays.

At the end of the crop, the house should be thoroughly cooked-out. Trays and woodwork should be treated with disinfectants before re-use. If only a few trays are affected, it is worth marking these and giving them a prolonged dip treatment in disinfectant.

ACTION POINTS

- Mark affected trays or areas of bed;
- Isolate the affected areas and treat them with 0.5% formalin;
- Pay special attention to the disinfection of all trays and areas of bed where symptoms have been seen; it is essential to kill all mushroom mycelium;
- Cook-out at the end of cropping;
- Examine peak-heat to make sure that some compost is not becoming excessively wet.

## Pit

Although this disease is fairly common, it rarely causes large crop losses. The cause of pit has not been established, although it is thought to result from attack by a bacterial pathogen. Mites and nematodes have also been implicated.

SYMPTOMS

Small, dark (often black), slimy pits appear on the cap surface of an otherwise healthy mushroom. The pits may penetrate the cap to a depth of several millimetres. Pits appear to be randomly distributed on the cap surface and varying in numbers from one to ten or more per mushroom. It is likely that the symptom is initiated at a very early stage in the development of the mushroom and shows up as a result of growth. The presence of such pits reduces the quality of the crop, often making it virtually unsaleable. Generally, pit does not appear until fairly late in the crop and most frequently is seen in the third or later flushes. It is often associated with conditions of very high relative humidity and poor evaporation.

CONTROL

As this disease is thought to be caused by a bacterium, those measures described for bacterial blotch (*see* page 76) are usually employed for its control, if the disease is sufficiently serious to justify any action.

ACTION POINTS

- Check peak-heat temperatures to make sure that these are high enough to kill pests such as mites or nematodes;
- Check hygiene during cropping, and also the cleanliness of the area used for storing and mixing casing ingredients.

## Drippy gill

This disease is caused by the bacterium *Pseudomonas agarici* and is of sporadic occurrence. It is most severe in the autumn and winter months.

SYMPTOMS

The gills are attacked by the bacterium even before the veil of the mushrooms is broken. Affected gills are often underdeveloped, showing small brown decaying areas with creamy-white bacterial ooze on them; hence the name, drippy gill.

Little is known about the conditions which favour this disease. However, it is likely that flies, pickers and water splash spread the organism within the crop. Because the gills are affected before the veil breaks, it seems likely that the bacteria are systemic within the mushroom. Infection may, therefore, take place at an early stage in the development of mushrooms, or the disease may originate from infected mycelium.

CONTROL

Little is known about the disease, and there are no specific control measures. Those recommended for bacterial blotch are usually employed.

ACTION POINTS

As for bacterial blotch (*see* page 78).

# 6

# Virus Diseases

## Introduction

Virus as a cause of disease of mushrooms was first described in 1950 and since then various diseases have been described. In the 1960s, many outbreaks were reported and considerable losses in crop yield were attributed to virus disease. More recently, with rapid diagnosis and a better understanding of the control, virus diseases have become more sporadic in occurrence, but can still cause considerable crop losses.

## The viruses involved

Various mushroom viruses (MV) are recognised, and these are differentiated according to the size and shape of the particles (*Figure 6.1*). In the United Kingdom, the accepted nomenclature for these are:

MV1: spherical particles, diameter 25 nm (1 nanometre (nm) = $10^{-9}$m or one millionth of a millimetre);
MV2: spherical particles, diameter 29 nm;
MV3: bacilliform particles, 50 x 19 nm;
MV4: spherical particles, diameter 35 nm;
MV5: spherical particles, diameter 50 nm.

In addition, particles of different shapes and sizes have been described in other countries, including club-shaped particles found in France, Germany and South Africa. Mushrooms are often affected simultaneously with more than one virus and, more often than not, 25 nm and 35 nm particles occur together. In the United Kingdom the frequency of the various mushroom viruses has changed with time, such that MV2 and MV5 are not now found (*Figure 6.2*).

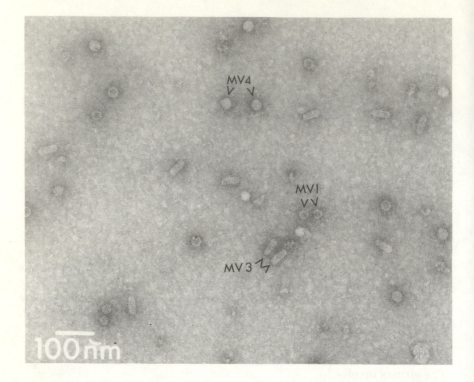

**Figure 6.1** Particles of mushroom virus 1 (25 nm spheres), mushroom virus 4 (35 nm spheres) and mushroom virus 3 (sausage-shape or bacilliform 50 nm x 19 nm). Crown Copyright

It is not known with certainty, or to what extent, each virus combination affects mushroom growth and yield. Experimental results indicate that MV1 and 2 can cause severe loss, and MV2 has also been associated with the dieback symptoms of mushroom mycelium. MV4 and MV1 are now the most common viruses in the U.K. crop and significant yield reduction is almost invariably associated with high concentrations of MV4; a return to low or nil levels coincides with improved cropping. The significance of MV3 and MV5 is unknown.

SYMPTOMS

Various symptoms may be attributable to virus disease, although the most consistent is a reduction in yield. None of the other symptoms are diagnostic.

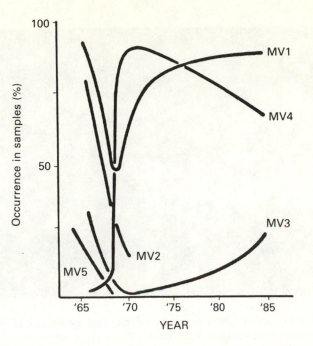

**Figure 6.2** The occurrence of mushroom viruses in samples tested at the Glasshouse Crops Research Institute (reproduced by kind permission of Dr R. J. Barton from his paper in the *Mushroom Journal* No. 151, 229–237, 1985)

When the attack is severe, crop loss can be considerable. Distortions of the sporophores have been reported, including elongation of the stalks, tilting of the caps and very small caps on normal-sized stalks *(Colour Plates 11 and 12)*. Affected crops may show a patchy appearance sometimes associated with death of the mushroom mycelium *(Figure 6.3)*. In other crops there may be an overall deterioration and reduction in cropping without very distinctive symptoms. Early maturity (premature opening) of sporophores, or slight discoloration of mushrooms are also symptoms, although these are not reliably diagnostic.

DETECTION OF MUSHROOM VIRUSES

Viruses cannot be detected using the techniques available for the isolation of fungal or bacterial pathogens: this is largely because virus particles are very small and cannot be seen under even the most powerful light microscopes; in addition, they cannot be cultured on an artificial medium. Various techniques

**Figure 6.3** Bare patches in a crop can be the first symptoms of virus attack. Apparently normal mushrooms from around the edge, when examined using electron microscopic techniques, show virus particles. Such bare patches can also result from other causes, e.g. uneven depth of casing, and are not a diagnostic symptom of virus attack

have been devised to overcome these difficulties. The techniques available are:

1. Comparative growth rates of mycelium on agar;
2. Direct electron-microscopic examination (EM);
3. Immunosorbent electron microscopy (IEM or ISEM);
4. Polyacrylamide gel electrophoresis (PAGE);
5. Enzyme-linked immunosorbent assay (ELISA).

It is worth considering the relative merits and possible uses of these.

*Agar growth test*

This was the first test to be devised and depends upon the fact that affected mushroom mycelium has a slower growth rate than otherwise identical healthy mycelium. For this test, it is first necessary to make a culture from the suspected sample of mushrooms and then to compare the growth rate of this culture with that of the identical but healthy mycelium of the same strain. It usually takes up to 2 weeks to obtain the initial culture and a further 2 weeks to complete the comparative growth test, provided that a culture of the healthy spawn is readily available. The advantage of this system is that it does not require expensive equipment. The disadvantages are, firstly, the long time needed to obtain a result and, secondly, the relative insensitivity of the test — although when growth differences are large, there is no doubt that a virus problem exists. In such cases, there is almost always a considerable crop loss. An arbitrary difference of at least 15% reduction in growth of the suspect sample has been used to define the presence of a virus disease.

## Electron microscopy (EM)

To examine mushroom samples by EM it is first necessary to make a suitable preparation: this is done by squeezing the juice from suspect sporophores, which is then filtered to remove large pieces of tissue. The filtered juice must be stained before examination and a heavy metal stain such as phosphotungsten is usually used; the stained juice is applied to a grid, which allows it to be examined. The operator must search the grid systematically for virus particles. They are easy to find when they are numerous. However, more frequently, when the virus particle concentration is low, it can take a long time to search and to be certain that no virus particles are present. One of the advantages of this technique is the speed in the detection in samples when large numbers of virus particles are present. The disadvantage is its uncertainty of detecting levels of virus too low to cause disease at the time of examination but which may indicate a potential problem.

## Polyacrylamide gel electrophoresis (PAGE)

This technique, known as PAGE, depends upon the detection of double stranded (ds) ribonucleic acid (RNA) — the infectious component of mushroom viruses. In order to use the technique, the virus RNA must be extracted from affected mushrooms. This can then be identified by applying the preparation to an agar gel column, which is subjected to an electric field. By staining the column after a predetermined time, the RNA, if present, can be identified. The advantage of this technique is its sensitivity, which is greater than that of direct EM examination (approximately 20 times more sensitive). It can also be used for identifying specific viruses. Its disadvantage is that it is dependent upon the stability of the virus particles during extraction of the ds RNA. Experience so far suggests that only MV1 and MV4 are sufficiently stable to be successfully extracted. The test takes from 1 to 3 days to complete.

*Immunosorbent electron microscopy (ISEM or IEM)*

In this test the electron microscope grids are coated with carbon, which strongly absorbs protein. The grids are floated on a drop of antiserum (containing protein antibodies) so that the antiserum is absorbed on to the carbon of the grid. The grid is then exposed to an extract of juice from an affected mushroom. Any virus particles present become attached to the antiserum on the grid. After staining, the grid is examined and the virus particles can be seen (*Figure 6.4*). The advantages of this method are its great sensitivity (5000 times more sensitive than EM), and also the fact that a relatively impure antiserum can be used and not one which is as specific as that needed for some tests, such as ELISA.

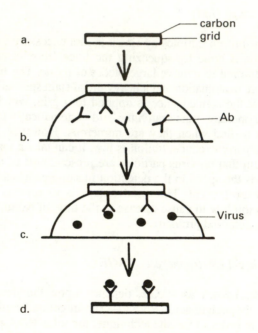

**Figure 6.4** Immunosorbent electron microscopy (ISEM): the procedure. (a) Carbon coating is applied to the electron microscope grids; (b) Coated surface of grid is floated on antibody (Ab) which becomes strongly bound to the carbon; (c) Antibody-coated grid is floated on mushroom extract and virus particles are allowed to be absorbed selectively by antibody; (d) Grid is stained and viewed in the electron microscope (reproduced by kind permission of Dr R. J. Barton from his paper in the *Mushroom Journal* No. 151, 229–237, 1985)

*Enzyme-linked immunosorbent assay (ELISA)*

The ELISA test requires pure preparations of the viruses in order to produce highly specific antisera. So far, it has proved to be very difficult to purify mushroom viruses to this degree and, currently, the test can be used only for MV3. The technique involves the attachment of a specific antiserum to a plastic well and the addition of a sample of juice from a suspect mushroom. If MV3 is present, it attaches to the antiserum. A further antiserum is then used with an enzyme attached to this preparation, which reacts with a substrate, causing a colour change to occur (*Figure 6.5*). Many tests can be done simultaneously using ELISA plates, and the sensitivity is about 500 times greater than the EM method. The main disadvantage of this technique is the necessity to have purified virus preparations but, if this can be overcome, the ELISA test is a simple and very effective method of virus detection.

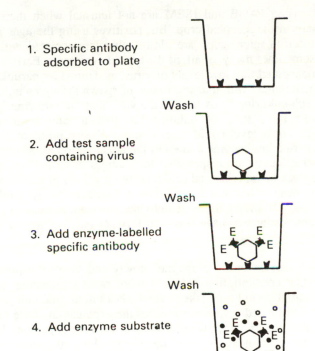

1. Specific antibody adsorbed to plate

Wash

2. Add test sample containing virus

Wash

3. Add enzyme-labelled specific antibody

Wash

4. Add enzyme substrate

Colour intensity proportional to virus concentration

**Figure 6.5** Enzyme-linked immunosorbent assay (ELISA): the principle (reproduced by kind permission of Dr M. F. Clark from his paper with A. N. Adams in the *Journal of General Virology* **34**, 475–483, 1977)

INTERPRETING THE RESULTS OF VIRUS TESTS

At present in the United Kingdom, the direct EM test or the ISEM test are most commonly used, the latter being very much more sensitive. Negative results by EM are often accompanied by positive results by ISEM. However, at present, positive EM results are much more likely to indicate a virus problem which is sufficiently severe to affect yield. Positive ISEM results, accompanied by negative EM results, are not associated with detectable effects on yield, but may serve as an early warning to the grower. It is not known what factors cause a low level of virus incidence to become a high level, with resultant crop loss.

LATENCY

Positive results using PAGE and ISEM are not unusual when there is no apparent problem in the sampled crop. But positives using the agar growth test or direct electron microscopy are almost always associated with yield reduction and sometimes many, or all, of the typical symptoms. Early work in the 1960's demonstrated that low levels of virus, as defined by particle numbers, could be maintained within agar cultures of spawn through a number of generations of sub-culturing. This lead to the view that the virus/mushroom-relationship was in a state of balance but that at any time, totally unpredictably, the virus level could change and a disease problem develop. This quiescent period has become known as the latent period, or as latency. It is not known what factors are responsible for changes in virus level. This is partly due to the lack of experimental evidence on the effects of known viruses on the host, because it has not been possible to infect easily mushroom mycelium with purified preparations of individual viruses. A number of possible explanations for latency have been put forward but two are perhaps of particular note:

1. Mushroom viruses generally propagate slowly and there is a prolonged period at a low concentration before a more rapid exponential growth phase. This initial period may take several weeks in normal cultural conditions, and the affected crop never reaches the exponential phase of virus replication. This theory depends upon virus in the near exponential phase being recycled to the next crop and thereby reaching its phase of rapid replication earlier in the life of the crop, i.e. by the first or second flushes.

2. Disease only occurs when a certain combination of virus components, different viruses, or triggering components are present at the same time. A point in favour of this theory is the known existence of two, almost identical in size, double-stranded RNA components in both mushroom viruses

1 and 4. Less is known about the other viruses, although it is thought that MV3 contains a single component of single-stranded RNA. As MV1 and MV4 often occur together, it is possible that all four of the components, two from each virus, are necessary before there is a measurable effect on the crop. Alternatively, yet another virus may be needed and, in this respect, it is interesting to note that club-shaped particles — associated with particularly severe problems — have been found in France, South Africa and Germany. These particles are very unstable and could easily be missed in most examinations. This theory also depends upon a change which results in the supply of the missing components necessary to trigger a disease problem. Other possible additional triggering components include pre DNA viruses and viroids.

Understanding the process of virus multiplication in mushrooms is essential for the interpretation of test results. ISEM or PAGE positive will mean nothing if the second of the above theories is the operative one. Conversely, if theory 1 accounts for the change, then the ISEM results could be a vital early warning, enabling growers to take avoiding action before the disease appears.

DISEASE EPIDEMIOLOGY

Work on viruses of fungi has so far demonstrated only two possible means of dissemination, from mycelium to mycelium through hyphal fusion (anastomosis) with the transfer of cell contents, and by transfer in spores which germinate and pass virus to healthy mycelium, again by anastomosis. As far as is known, there are no insects, nematodes or other vectors of mushroom viruses, although it is, of course, known that virus infected mushroom spores can be distributed by flies and other insects. Changes in the frequency with which the different mushroom viruses have been detected over the years is thought to reflect, at least in part, their means of transmission.

Mushroom viruses 1, 3 and 4 are readily spread in mushroom spores and this is probably the most important means of dissemination. It is believed that MV2 is transported only ineffectively in this way, but, like the other viruses, is also readily transmitted through infected mycelium.

Mushroom spores become airborne easily, but they can also be spread by workers and probably by some flies, such as sciarids. Distant spread can occur, although the most likely source of mushroom spores and, therefore, of virus particles, is the mushroom crop. Wild mushrooms are not proven hosts of the viruses which affect *A. bisporus*. Spores are stimulated to germinate by the presence of mushroom mycelium. Infection of the mycelium at spawning or during spawn-running and certainly before casing, is likely to result in a maximum reduction in yield. The compost may also become contaminated by infected mushroom spores during cool-down after peak-heat. Once the virus has become established within the crop, the particles are readily spread through the mycelium.

Mushroom viruses can also survive in mycelium which is left attached to trays at the end of the crop. If cook-out is not effective, the virus will remain in the viable mycelium and will be transferred to the new crop following hyphal fusion between the old and new mycelium. Once the mycelium is dead, the virus particles become non-viable.

Mushroom spawn is obviously a potential source of viruses. Some reports suggest that viruses do not occur in spawn, whereas others indicate widespread occurrence. The general absence of virus disease problems from most crops points to a low level of incidence or absence in spawns, although it is possible that very low levels of virus remain undetected. The circumstances in which insignificant concentrations of virus change to significant and damaging ones are unknown.

*Spores as a source of virus*

Mushroom spores measure 5 x 7 μm and are produced in very large numbers in mushroom crops. It has been estimates that a single mushroom with an 8 cm diameter cap will produce 1,300 million spores. The numbers of spores in the air of a mushroom house has been measured at 10,000 spores per cubic metre, and even in areas where filtration is in practice, spore counts of up to 10 per cubic metre have been reported. Calculations show that as few as 10 spores per cubic metre can result in 10,000 to 100,000 landing on each tray during a normal 14 day spawn run. In experiments, as few as 10–100 viruliferous spores per tray will induce a recognizable disease problem.

Mushroom spores are stimulated to germinate by nearby mushroom mycelium. Some other fungi, such as *Peziza ostracoderma*, have been shown to exert the same stimulation. It is clearly very important to be able to eliminate mushroom spores as far as is possible. Their viability has been tested on a number of occasions. Stored at room temperature in a dry state, virus infected spores were shown to be capable of transferring virus to healthy mycelium after a period of at least 6 years. Stored at 4°C, mushroom spores were still viable after 10 years.

Experiments to determine the thermal death point of mushroom spores have not been exhaustive. Work with healthy spores showed that, treated either wet or dry, they could withstand 16 hours at 60°C, but not at 65°C. Treatment times of less than 16 hours were not tested. Similarly, virus infected spores were shown to be capable of virus transfer when treated in a peak-heat which reached a temperature of 60°C. Conversely, other workers have shown that temperatures as low as 54°C, maintained for 10 minutes, will kill mushroom spores.

Work on the sensitivity of mushroom spores to chemicals is even more sparse, but there are some indications that they might be able to withstand at least some of the chemicals used as disinfectants on most farms, even at the concentrations recommended.

## Mycelium as a source of virus

Viable mushroom mycelium must also be considered to be a very important source of mushroom viruses. Fragments left on or in woodwork and on netting could anastomose with new mycelium and transfer virus to the new crop. In the process of bulk spawn-running, and at any time when spawn-run compost is moved, it is possible that small fragments of mushroom mycelium become airborne, and may then be disseminated in much the same way as mushroom spores. Fragments less than one cell in size will not survive and it may be necessary to have mycelial pieces of at least several cells in length for survival. Such fragments will be considerably larger than mushroom spores and are unlikely to pass through filtration equipment, but may be present in filtered air areas where bulk run compost is handled.

## Anastomosis

Virus transfer from germinating spores or from mycelium depends upon anastomosis. How frequently does anastomosis occur? Within any one crop it is likely to be extremely common and any cultural operation which results in mycelial fragmentation, such as the movement of bulk run compost, ruffling, and spawned casing, will be quickly followed by large scale anastomosis. Anastomosis between strains may be commonplace or infrequent, depending upon the relationship of the strains. Those strains most closely related, e.g. smooth whites, will readily anastomose. Experimental work indicates that off-white cave strains do not readily anastomose with smooth-whites. Little is known about the anastomosis pattern of hybrid strains, either with smooth whites or off-whites. There is evidence that some smooth-white strains do not anastomose as readily as others. Advice should be sought from the spawn suppliers on the most suitable alternative strains. Anastomosis between different species of agarics would appear to be far less likely to occur.

One significant feature in the epidemiology of mushroom virus diseases in the last few years has been a close association between the occurrence of severe disease and of a period when mushrooms were not harvested on time, and many open mushrooms resulted. One consequence of this is to increase the chance of mushroom spores, possibly carrying virus, reaching new crops in larger quantities than normal.

By inoculation experiments with known infected mushroom spores, it has been clearly demonstrated that the most vulnerable stage of the crop is between cool-down of phase II and casing. Experimentally, infected spores applied within 10 days of casing affected yield to some extent but there was a decreasing effect with time from casing. Clearly, the stage in crop development to concentrate upon when considering control is between cool-down and casing.

CONTROL

Successful control of mushroom virus diseases is achieved by a combination of all the processes that constitute a thorough hygiene programme. If virus disease occurs, it is necessary to re-examine the whole farm programme in order to detect weaknesses. The aim of the programme for virus disease control is to ensure that no mushroom spores or viable mushroom mycelium come in contact with the new crop. The most vulnerable stage is between cool-down of the compost at the end of phase II and casing. There is then a reduced risk until first pick and, subsequently, the risk diminishes almost completely. If each of the cultural practices are examined throughout the production of the crop, the vulnerable points can be identified and control measures taken.

A summary of the aims and actions for a suitable hygiene programme to control virus are given in *Table 3.3* (page 38–41). At almost all stages in crop production there are necessary actions required to minimize or eliminate the chance of virus being introduced into a new crop.

*Phase I*

Virus infected spores from cropping houses deposited on a compost stack may survive the composting process. Those that are eventually 'turned' into the stack will be killed by the temperature or be composted, but some will remain on the outside, or may be deposited after the last turn. These spores, if they survive phase II, will constitute a source of virus. Siting the stacks away from the cropping houses, or fitting exhaust filters on the cropping houses, can reduce this risk if it is seen to be high. The correct siting of the yard in relation to wind direction, or the use of a windbreak, can minimize or eliminate the risk of stack contamination.

*Phase II*

There is conflicting evidence on the thermal death point of mushroom spores, varying from a complete kill at 54°C for 10 minutes to 65°C for 16 hours (in the latter case, shorter times were not tested but spores did survive 60°C for 16 hours). Experience in practice clearly indicates that kill temperatures of

60°C are effective in helping to control virus disease. If this is so, 60°C must be maintained throughout the compost for several hours in order to be certain that this potential source of the pathogen is eliminated. After the kill temperature has been reached and during the cool down period, newly introduced spores may be able to withstand the compost temperature. It is, therefore, very important that any fresh air is filtered. Generally, absolute filtration is used (*see* pages 26–27), which is designed to remove all particles and spores of 2 μm diameter or larger. Very few, if any, mushroom spores should get through such filters. Filtration of this type is even more critical when considering the bulk preparation of compost either in phase II or spawn-running. There is obviously little point in having expensive filtration if the phase II room is not air-tight. Holes in the ducting, gaps around the filtration unit, cracks in walls, badly fitting doors, can all contribute to the introduction of unfiltered air into a phase II. To some extent, small cracks and holes in some parts of the structure can be counteracted by the use of a positive air pressure within the room so that air is always flowing out and never in, except through the filtration unit. Other means of cleaning the air intake into phase II, such as by ultra-violet light treatment, have not been exploited in mushroom growing but may have application, particularly in the bulk preparation of compost.

*Spawning*

This is a very high risk stage for the introduction of various pathogens, as well as virus. It is important that the equipment used is thoroughly cleaned immediately before use and not the day before. This is best done by spraying with a disinfectant or, if the spawning equipment is within a sealed building, by fumigating. It is not necessary to use high volumes of water to disinfect, although a thorough washdown with a pressure hose may already have been done at the end of the previous process. A sprayer capable of delivering fine droplets at a high pressure will give all surfaces as good a cover as possible. Formalin used at 2% is suitable for this operation. It is very important to make certain that spawn-hoppers and supplement-hoppers are also thoroughly cleaned. The surrounding area should be thoroughly washed down with a disinfectant and, if it is a sealed area, this can be done the night before using a phenolic material, or formaldehyde. If the area is open, it is better to complete the washing-down as near to the time of spawning as possible, which may preclude formaldehyde from the choice of materials. Handling the spawn must be done with clean hands. Workers on the spawning-line should have clean overalls for the purpose. Overalls can be cleaned of all fungal spore contamination by running them in a tumble drier for 30 minutes set at the warmest setting, or by machine washing in the normal way.

*Spawn-running*

All the same conditions as apply to phase II are relevant here. Filtered air must be used, or air can be recirculated within an enclosed system with a cooler installed to regulate the air and compost temperatures. Spawn-running rooms and ducting must be thoroughly disinfected and/or fumigated before use. During spawn-running, the surface of the compost can be covered with paper or polythene in order to prevent contamination by fungal spores. Paper is often preferred because it absorbs moisture and prevents the surface of the compost becoming over wet. The paper should be sprayed with 2% formalin at 3 or 4 day intervals in order to kill spores landing on its surface. The last spray is best done just before paper removal to prevent spores transferring to the compost surface and with dichlorophen to avoid the unpleasant fumes of the formalin being inhaled by the workers removing the paper.

*Casing*

It is very important that the casing ingredients are free from contamination. Peat stored as wrapped bales and bags of chalk are likely to be 'clean', but bulk stored materials are subject to contamination by the settling of airborne spores in the same way as compost. It is, therefore, important to store casing materials in a separate area and to fumigate with formaldehyde at intervals to ensure cleanliness.

*Post casing*

The crop is much less vulnerable at this stage, but two recent developments may be relevant. It is now common practice to ruffle the casing when the mushroom mycelium has grown into it in order to regulate yield and quality (*see* page 7). This process is done either by hand using a rake, or by machine. Mushroom mycelium is broken and, inevitably, there is some movement of mycelial fragments along the bed. If small patches of virus are present, it is likely that affected mycelium will be spread. In order to prevent bed-to-bed spread, it is important to sterilize the raking equipment at regular intervals. This is best done by spraying with a disinfectant at the end of each bed. A sprayer with a fine droplet is ideal for this purpose and 2% formalin is an effective disinfectant. The practice of adding mycelium to the casing (spawned casing — *see* page 6) may also increase the risks. Obviously, the added mycelium must be free from virus disease but, if the crop is already infected, it is possible that additional spread will occur during the process. Cleaning of equipment in a way similar to that already described for ruffling should be practised.

*Cropping*

If the farm policy is to grow flats, or because of picking problems it has become unavoidable, mushroom spores will be produced in large numbers. An exhaust air filter (a 2 cm thick layer of fibre glass matting is adequate), with the element changed regularly, will help to reduce the spore load in the exhaust air. Such filtration should be a matter of course if flats are continuously grown.

At the end of the crop, cooking-out *in situ* at 70°C for 12 hours will ensure that all mushroom spores and mycelium are killed. A much shorter treatment time at this temperature is probably effective but, in order to be certain that all parts of the compost, the woodwork, and structure reaches a high enough temperature, it is generally necessary to treat for a 12 hour period.

If cook-out is done in a special room and trays are removed from the cropping houses before treatment, they must be thoroughly drenched with a disinfectant shortly before removal in order to kill all the spores on their surfaces. A phenolic material is suitable for this purpose. Where the crop is growing on shelves it is important to thoroughly clean the matting before re-use. Equipment is available that will allow this to be done. The metal structure should be thoroughly washed and sprayed with 2% formalin before the cleaned matting is returned.

*Tray cleaning*

Even after cooking-out it is advisable to treat trays and, if cooking-out has not been possible, then tray cleaning is essential. After emptying, trays should be pressure washed in order to remove compost. They can then be immersed in a disinfectant such as sodium pentachlorophenoxide (Cryptogil) or dichlorophen (Panacide M). Maintaining the correct strength of the disinfectant in the dip tank is vitally important. The trays are drained and allowed to dry and, if treated with Cryptogil, they may be washed with clean water after treatment in order to remove some of the chemical so as to avoid discolouration of the mushrooms at the edges of the boxes. This is particularly important where watering trees are used, which tend to splash disinfectants from the wood onto the crop. After treatment, the trays should not be stored in an exposed position where they may become contaminated with mushroom spores. If this is unavoidable, they should be sprayed with dichlorophen immediately before filling.

Finally, washing down the farm at least once a week with a disinfectant ensures that mushroom spores on roadways are killed. Similar treatment should be given to vehicles and tools.

*Strains*

Some strains of *A. bisporus* do not show symptoms as markedly as others. These are the brown, cream and off-white types. The related species, *A. bitorquis,* does appear to be tolerant, or perhaps resistant, to the viruses which affect *A. bisporus.* A complete cropping cycle around the farm of off-white strains, or some smooth-white strains known to anastomose less frequently with others, or of *A. bitorquis,* can help to reduce the general virus inoculum and can enable economically worthwhile crops to be grown.

Off-white, cream or brown strains of *A. bisporus* and strains of *A. bitorquis* do not readily anastomose with white strains of *A. bisporus,* so the chance of virus transfer is greatly reduced. Hybrid strains (hybrids between white and off-white types of *A. bisporus*) can anastomose with both white and off-white strains and, therefore, their widespread culture may reduce the effectiveness of strain alternation as a means of virus control.

ACTION POINTS

- Pay strict attention to hygiene at all phases of crop production to help to ensure that virus diseases do not become a major problem;
- Filter the air during peak-heat and spawn-running;
- Change to low risk smooth-white (consult your spawn supplier), an off-white or cream spawn, or to *A. bitorquis* for a period to enable the inoculum level on the farm to reduce;
- Beware that, where hybrid strains are used, the change to non-hybrid white or cream strains may not be effective;
- Whenever possible, pick all mushrooms before they have opened in order to prevent the dispersal of spores.

# 7

# Moulds in Mushroom Crops

## Introduction

Various moulds (*see Glossary*) are associated with mushroom growing and they are often named according to the colour of the spores or mycelium seen in the crop. They occur chiefly when the compost has not been properly prepared and the growing conditions are not favourable for mushroom growth. Moulds are generally considered to be competitors for nutrients, or antagonists, rather than parasites. Some of the most common are described below. Control measures are similar for most of them and are given at the end of the descriptions (page 109); specific points for each mould (where relevant) are included with the descriptions.

## *Penicillium* mould

Various species of *Penicillium* are associated with mushroom culture and, generally, they do not cause problems. The characteristic green mould is commonly seen growing on trays or the sideboards of shelves, or pieces of sporophore tissue left on the casing surface, and on grains of spawn. Apart from the appearance and some inconvenience, this mould has no measurable effect on crop yield or quality. However, recently there has been a number of serious cropping problems in the United Kingdom associated with one species.

## Smoky mould (*Penicillium chermesinum*)

This species has not previously been recorded from mushroom culture, and there are very few international records of its occurrence in any situation. It is thought that, in the current outbreaks, the mould may have originated from dirty straw. It has not been found on wood or other materials within crops.

The *P. chermesinum* problem is characterized by a dramatic reduction in yield which, at the extreme, can be 80%. Symptoms become apparent in the first flush where there is an edge break of more or less normal mushrooms, with no cropping in the middle of the shelves or boxes. If the compost is examined in the non-cropping areas, large clouds of spores looking like smoke can be seen as the compost is disturbed. The compost has a mouldy smell. In less severe outbreaks, the first and second flushes may be reduced in yield and, by the third flush, non-cropping areas occur, together with the characteristic spore production from the compost. When the affected compost is examined very carefully, numerous penicillate sporing structures are seen. These are white in colour but, with age, turn brown. On agar, the fungus has a very characteristic slow growth rate, and on potato dextrose agar, is grey-brown in colour.

The spores of *P. chermesinum* are slightly oval in shape and are just under 2 μm at their narrowest point. It is, therefore, possible that some could pass through a 2 μm filter.

This weed mould problem has, so far, mainy been associated with farms where the compost is bulk handled. Examination of compost in bulk phase II has shown the presence of the spores in the layers against the netting. When this compost is moved, the *Penicillium* spores are distributed throughout the compost.

It is likely that there is an association between *P. chermesinum* and mushroom mycelium, perhaps similar to that already reported for *Trichoderma harzianum*. When maximum damage occurs, it seems highly likely that the *P. chermesinum* spores must be present in large quantities within the compost at the time of spawning and the mushroom mycelium is either parasitized or very potently inhibited. Bulk phase II and/or bulk spawn-running systems are ideal for spore distribution and, as the *Penicillium* has an optimum growth temperature of 28°C, it is ideally adapted to compete with, and perhaps attack, the mushroom mycelium during the initial stages of mycelial colonization.

Controlling this problem has proved to be extremely difficult, particularly in situations where there is a continuous flow of bulk compost within an area where spawning also occurs.

Filtration of the air used in phase II is obviously vitally important and any small cracks or gaps within the system will result in compost contamination. Filters must be carefully checked to make certain that they are efficient and replaced at the correct intervals.

Areas outside the bulk tunnels must be regularly disinfected and, if within a sealed spawning area, fumigated regularly with formalin to make certain that airborne mycelial fragments and spores that have settled on ledges are killed. It is particularly important to do this every time the tunnels are emptied, whether they are for phase II or bulk spawn-runs.

Some experimental work has shown that benomyl added to the compost at spawning gives an effective control, but this procedure should not be considered as a routine as it is highly likely to result in fungicidal resistance.

The procedures described for virus disease control should be followed carefully when considering the control of the *Penicillium* problem as the crop is likely to be most vulnerable at the same stages of crop production.

## *Pythium* – black compost

At the end of the spawn-run, compost may show black areas of totally uncolonized compost when the adjacent areas have been satisfactorily colonized. The black patches may vary in size from a few centimetres to a metre or more in diameter. They are usually just below the compost surface and may not be at all apparent at casing, unless the compost is examined carefully. Also, they are most frequently found in the top half of trays, although they can extend deeper. The patches may, in fact, be ovoid in shape, with the upper part of the affected compost just touching the surface.

Examination of such compost in the laboratory shows it to be colonized by a species of *Pythium*. In the United Kingdom, the species involved is *P. oligandrum*, a well-known pathogen and antagonist of various fungi. In the U.S.A., *P. hydnosporum* (syn. *P. artotrogus*) is reported as the cause of the problem.

*Pythium* resting spores are very resistant to heat and drought and have been recovered in a viable state from dry compost on the surface following phase II. In experiments, the maximum inhibitory effect was shown to occur when the *P. oligandrum* was introduced either before spawning or with the spawn. There was no effect on spawn growth if the *Pythium* was introduced two days after spawning.

This work suggests that, for a problem to occur, the compost must be contaminated before spawning. It seems that such a soil-borne fungus is most likely to be a contaminant of the straw either in soil splashed on to the straw during crop growth, or by storing the straw bales on soil before use.

Another association in this problem is that of high nitrogen in the areas where the *Pythium* occurs. In one case at least, this has been associated with the uneven distribution of a nitrogen supplement at spawning. It is not clear whether the areas of high nitrogen prevent mushroom mycelial growth, or whether they encourage the growth of *Pythium*.

Generally, crops grown from such affected compost are reduced in yield especially in the first and second flushes but they may make this up, to some extent, later. Eventually, the blackened areas become colonized by mushroom mycelium, perhaps when the high nitrogen levels are reduced by leaching.

To control this problem, it is important to avoid the necessary contamination of straw with soil. Also, care should be taken when supplementing to make certain that the supplement is as evenly spread as possible over the compost surface, thereby avoiding patches of excessively high nitrogen concentrations. *Pythium* spp. are generally encouraged by wet conditions, so a re-examination of the moisture content of the compost at spawning may be worthwhile.

## Olive-green mould (*Chaetomium olivaceum*)

Mycelium of this fungus is grey-white and is often mistaken for mushroom mycelium during the early stages of spawn-running. It is a fairly common mould and, when extensive in compost, reduces yield in proportion to the extent of the colonization of the compost. It does not grow in casing. The fungus is usually identified by the olive-green fruiting structures which are about the size of a pin-head and with a rough, spiny appearance. These are produced in large numbers on the straw, often well spaced, and clearly seen with the naked eye. As they age, they darken in colour, eventually becoming brown *(Colour Plate 13)*.

Compost with olive-green mould is often black and is not colonized by mushroom mycelium. Some mushroom growth may take place into the affected areas but only sparsely and not enough to give a satisfactory crop. If extensive olive-green mould develops shortly after the end of peak-heating, there is little hope of achieving a satisfactory yield.

Olive-green mould development is favoured by anaerobic conditions during peak-heat. Such conditions may occur when the compost is too wet, over-composted, overheated (a compost temperature greater than 62°C (143°F)), not adequately aerated during peak-heat, or over-compacted at filling. *C. olivaceum* is able to withstand higher levels of ammonia than the mushroom mycelium, so thrives in conditions that are adverse for spawn growth.

Good composting in phase I and control of the environment during phase II will prevent olive-green mould but there is little that can be done to save an affected compost. Re-heating or re-spawning such a compost is not successful.

## Fire mould (*Neurospora crassa*)

This fungus commonly occurs after cook-out, particularly when the crop is left *in situ* after treatment. The mycelium is at first creamy white but rapidly turns orange. Large wefts of mycelium form, resembling cotton wool, often hanging down like cobwebs. Large numbers of spores are produced so that, once the mould is established on a farm, it is difficult to eliminate. There are no specific control measures.

### Red geotrichum or lipstick mould (*Sporendonema purpurescens*)

The fungus produces a fine white mycelial growth on the surface of the casing or in the compost, not unlike mushroom mycelium. Its colour generally changes to bright pink and finally buff, with a powdery appearance as the spores are produced. In peat and chalk casing, the characteristic pink colour does not always develop, the colonies remaining white. The large numbers of spores easily become airborne. The presence of this mould in the crop has a marked inhibitory effect on the growth of mushroom mycelium and is often associated with crop loss, particularly if it is present in the compost before casing. Raising the temperature in peak-heat to 65°C (149°F) for 4 hours is said to eliminate lipstick mould from the compost but this procedure may increase the chances of other weed moulds developing. Generally, strict attention to hygiene is the most effective and safest approach to control.

### Sepedonium yellow mould (*Sepedonium* spp.)

This mould develops in the compost and, although initially white, turns yellow to tan with age. It produces numerous spores which become airborne easily and can recontaminate compost during preparation. The large spherical spores of this fungus are said to be very heat resistant. It has recently been found in the basal layers of compost produced by bulk methods, and also in the bottom of troughs. Its presence in troughs has been associated with distortion of mushrooms, possibly due to the production of a volatile toxin.

*Sepedonium* affects mushroom growth in compost considered to be suitable for mushroom mycelium. It is said to be associated with a yield reduction, but this has not been quantified. Its association with the lower layers of bulk produced compost has been linked with wetness.

### Mat and confetti (*Chrysosporium luteum* and *C. sulfureum*)

These are both yellow moulds that were common but are now less frequently found. Where they occur in large quantities, considerable yield reduction may result. Most serious losses are associated with the occurrence of these moulds very early in cropping.

With mat disease, a yellow-to-brown mycelial layer forms at the casing-compost junction. Initially, the colour of the mycelium is very similar to that of mushroom mycelium and is usually only distinguishable after spawning, when the mycelial mats bind the compost together.

Confetti disease develops in the compost and the yellow colonies of the fungus are produced from an initial growth of white mycelium. The colonies may coalesce to form large, dense yellow patches. Spores of both these fungi become airborne and will cause trouble if they contaminate casing or compost. Soil is a common source of both fungi and cleaning the composting area, to make certain soil water or soil does not contaminate the compost, will help to control these moulds. In extreme cases, filtration of the phase II may also become essential.

### Brown plaster mould (*Papulaspora byssina*)

Brown plaster mould was once a very serious problem but, with improved standards of peak-heat, it is now less frequently seen. Characteristically, it produces large, dense, roughly circular patches of mycelium on the surface of the casing, initially whitish, but turning brown and powdery with age (*Figure 7.1*); it can also colonize the compost. The fungus produces numerous spores which consist of clusters of cells, often referred to as bulbils. The presence of this fungus has been associated with wet compost.

**Figure 7.1** Brown plaster mould (*Papulaspora byssina*) growing over the surface of casing

**Figure 7.2** Small, dark brown, gelatinous, disc-shaped apothecia of *Peziza ostraco-derma*, seen here on the surface of a plant pot, but they commonly occur on mushroom beds

## Cinnamon mould (*Peziza ostracoderma* syn. *Plicaria fulva*)

This is one of the commonest of the brown mould which frequently occurs on the surface of the casing before the first flush. The colonies are often circular and, initially, are grey-white but quickly turn brown. By the time that the first flush of mushrooms have developed, the cinnamon mould has often disappeared. Two spore forms are produced: numerous small conidia, which are very readily distributed from the brown patches and ascospores, which are produced by small, dark-brown, gelatinous, disc-shaped, circular structures (apothecia) about 1 cm across (*Figure 7.2*). These latter structures usually appear well after the obvious mould growth has gone.

Cinnamon mould is more of a nuisance than a cause of crop loss, although, where it occurs in quantity, it depletes the nutrient status of the compost, may disfigure the mushrooms, and can result in a small delay in cropping, with a possible reduction in yield.

### Brown mould (*Arthrobotrys* spp.)

*Arthrobotrys* spp. also produce brown colonies on the casing surface, not unlike those of cinnamon mould in appearance, except that they occur at any stage in cropping and tend to be more common at the end of the crop. They are parasites of nematodes, and their presence is an indication that the nematode population of the compost and/or the casing is high. A preparation of this fungus has been sold in France to add to compost or casing as a means of improving yield, possibly by attacking nematodes.

### White plaster mould (*Scopulariopsis fimicola*)

This fungus produces dense white patches of mycelium and spores on the casing surface and in the compost. Many other weed moulds initially produce white mycelium but they change colour as they age, whereas *S. fimicola* remains white. The fungal growth is sometimes so dense that it looks as though a bag of flour has been emptied on to the bed surface. White plaster mould is now uncommon but, when present, can reduce yield by competing with mushroom mycelium. It grows particularly well in compost with a pH of 8.2 or above, a condition which often results from under-composting.

### Black whisker mould (*Doratomyces stemonitis*)

This mould is so named because of the dark-grey to black whisker-like bristles produced on the surface of the casing. These bristles are spore-bearing structures of the fungus and may be up to 2 mm long. *Doratomyces* is said to indicate poor initial composting, which has resulted in an abnormally high cellulose content of the compost. The fungus is thought to be antagonistic to mushroom mycelium and may affect yield. It is frequently associated with *Penicillium* and *Aspergillus* spp. and some pickers are said to be allergic to the spores. The full significance of this fungus is not understood.

### Ink caps (*Coprinus* spp.)

The *Coprinus* fruit bodies (*Figure 7.3a,b*) are produced on mushroom beds, usually before cropping begins and these subsequently disintegrate into a black slime (*Figure 7.3c*). The mycelium of the fungus is grey and not easily distinguishable from mushroom mycelium. Ink caps often occur in large numbers before the first flush and, in spite of their presence, the mushroom crop often grows well. The large numbers of spores released from the *Coprinus* sporophores will readily colonize freshly prepared compost. The presence of ink caps can indicate free ammonia or, ultimately, a high nitrogen status of the compost, and the yield of mushrooms may subsequently be satisfactory. However, ink caps also grow well if peak-heating has been inadequate and are then, conversely, associated with very poor yields. Generally, ink caps do not occur if compost preparation has been satisfactory.

**Figure 7.3a** Ink caps (*Coprinus* spp.) developing in a mushroom crop, the young fruiting bodies growing in a patch of cinnamon brown mound (*Peziza ostracoderma*)

**Figure 7.3b** *Coprinus* spp.: mature sporophores

**Figure 7.3c** *Coprinus* spp.: shortly after maturation sporophores dissolve (deliquesce) and rapidly disappear

ACTION POINTS FOR CONTROL OF MUSHROOM MOULDS

- Check the peak-heat temperatures and, in particular, the variability in compost temperatures in different parts of the peak-heat room;
- Check the ammonia level in the compost at the end of peak-heat;
- Check hygiene measures in the compost area and the casing mixing and storing areas;
- Make sure all machinery is thoroughly cleaned after use;
- Check the cook-out temperatures, making sure that 70°C (158°F) is achieved throughout and maintained for 2 hours;
- If methyl bromide is being used, ask the contractor to recheck his rates of use and the concentration achieved during treatment, as well as its duration. The correct temperature at the time of treatment is also important;

- In addition to cook-out, make sure all boxes are thoroughly disinfected between crops;
- Install or check filtration in the phase II and spawn-running areas;
- Check compost during spawn-running for the correct environmental condition, both in the compost and in the atmosphere;
- Where problems are severe, follow the recommendations for the control of virus diseases (page 94) and *Trichoderma harzianum* (page 67).

# 8

# Pests

## Introduction

Mushrooms are grown on a continuous cycle throughout the year, in a protected environment of warm, constant temperatures and high humidities. The crops grown throughout the year, although possibly of varying strains, are, with few exceptions, a monoculture of *A. bisporus*. It is not surprising, therefore, that commercial mushroom production is affected by a number of pests capitalizing on the bountiful and continuous supply of nutrients available to them.

The fauna associated with phase I composting is extensive, but will not survive an efficient peak-heat. It follows, therefore, that for subsequent crop loss to occur, pests must colonize a crop after this pasteurization process. As the initial infestation level of any flying pest is likely to be low enough to escape casual observation, it is good management practice to use some sort of monitoring technique to determine the relative abundance of such pests. Simple sticky traps, which are commercially available, can be used to good effect, either with or without a small adjacent light source.

When a potential pest is first noticed, it is extremely important to identify it correctly, as control regimes for the various pests differ in enough respects to cause control failures if a mis-identification occurs.

A list of approved insecticides suitable for the control of some insect pests of mushrooms is given in *Table 8.1* (pages 128–129).

## Mushroom flies

The most serious pests of cultivated mushrooms belong to a group of insects called the Diptera, so-called because the adults are two-winged flies. As the various pest species are similar in size and colour, they are often referred to as simply 'flies', a general term which belies the importance of accurate identification for sound control practices.

111

## Sciarids

Several species of sciarid have been recorded as infesting mushroom crops in Britain, although intrinsic difficulties in sciarid identification have led to some confusions in the past. Only two species could have been regarded as being common — *Lycoriella solani* and *L. auripila* — with the former being only rarely seen in recent years. *L. solani* used to be the dominant species, but in the last 12–13 years *L. auripila* has become the most important species in this country. It causes more primary damage to mushrooms than *L. solani* and, since developing resistance to organophosphorous (OP) insecticides such as diazinon and chlorfenvinphos, it has become the most damaging mushroom pest. Information gathered from trapping studies indicates that sciarids tend to stay on a farm all year round, in contrast to phorids, which are more seasonal pests, a factor that may have contributed to the development of insecticide resistance.

SYMPTOMS

Sciarids are able to affect all stages of post peak-heat mushroom production. A heavy infestation by flies at, or just before, spawning may inhibit the spawn-run as the larvae, developing and feeding within the compost, will produce large quantities of faecal matter, and break down localized areas of compost into a soggy, foul mess. The mushroom mycelium will be unable to colonize this contaminated material and, thus, poor yields will subsequently result.

The most often reported damage caused by sciarid larvae is tunnelling in the stipes, as the presence of tunnels in a cut mushroom is easily noticed *(Figure 8.1)*. Stipe tunnelling appears to occur only when the population of sciarids is high but, when it does occur, it can be quite extensive. However, the most serious damage, in economic terms, is the attack by larvae on the developing pin-heads and buttons. These structures are very much smaller than a mature mushroom and, consequently, are generally unable to survive a larval attack. The mycelial attachments can be severed, causing the pin-heads to become brown and leathery; or the pin-head/button can be hollowed out (producing a sponge-like mass); or even consumed entirely. This sort of damage is insidious because it occurs at a lower population density than is –required for stipe tunnelling and can go unnoticed by the grower.

The adult flies are also capable of causing damage to a crop, although not in a direct manner. Various mite species are able to breed in the foul areas of compost and decaying pin-heads created by sciarid larvae. The mites are able to cling on to adult sciarid flies, and as many as 30 per fly have been recorded. As these mites are often associated with various bacterial diseases and *Verticillium*, their dispersal around a mushroom farm may cause undesirable

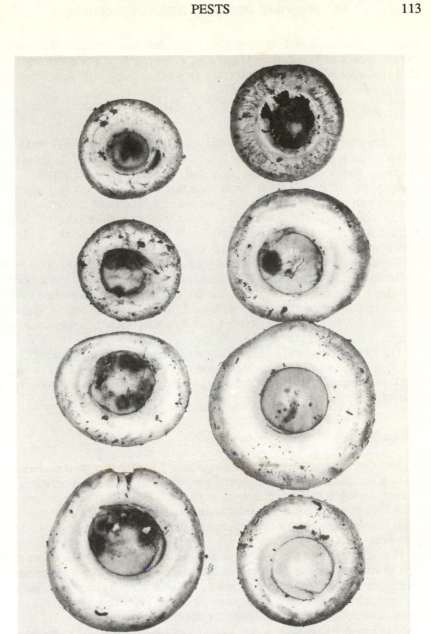

**Figure 8.1** A range of damage symptoms caused by larvae of the mushroom sciarid, *Lycoriella auripila*, burrowing within mushroom stipes

pathogen transmission. In a similar way, the adult flies are capable of spreading the spores of *Verticillium (Figure 8.2)*. They will, if not controlled, also act as a source of re-infestation for new crops.

DESCRIPTION OF PEST

Mushroom sciarids, sometimes called fungus gnats, are small (3–4 mm long), delicate, black gnat-like flies with large compound eyes and long thread-like antennae, which are held characteristically erect (*Figure 8.2* and *8.3*). The abdomen of the female fly is larger than that of the male (as it is generally bloated with eggs), and its apex is quite pointed. The male abdomen is quite slender and terminates in prominent genitalia with well-developed claspers (*Figure 8.4*). The characteristic venation in the iridescent wings is also an important identification feature of sciarids (*Figure 8.5*).

The adult flies are not as active as phorids near lights. Females tend to rest on the surface of trays and walls, while males often remain on the surface of the casing, in readiness to mate with the newly emerging females. In their natural habitat, sciarids inhabit leaf mould, wild fungi and rotting vegetable matter, and it is from these sources that they originally infested mushroom farms.

The larvae are white, legless, fairly active maggots ranging from 1–8 mm in length. The main identification feature is the distinct, large head, which is black and shiny (*Figure 8.6*) and bears large, powerful chewing mouthparts (*Figure 8.7*).

EPIDEMIOLOGY

Sciarids that infest the compost during phase I composting will be killed by the high temperatures reached during the peak-heat process. Populations that emerge during cropping, therefore, must be the result of infestations that occur after this process. These infestations arise from sciarids that are attracted to the fermentation odours being given off during the cool-down period of the peak-heat.

Females can lay up to 140 eggs and, as compost temperatures are quite high during the subsequent spawn-running period, the new generation of adults can emerge from the compost 2–3 weeks later. By this time, the crop has been cased and, in general, it is in the casing layer that subsequent generations develop and cause damage to the developing crop.

Experiments have shown that loss of yield is proportional to the average number of larvae present throughout the cropping period. A mean of just one larva in a handful of casing causes 0.5% loss in total yield. As the cost of the recommended control measure for sciarids (30 ppm diflubenzuron in the casing; *see Table 8.1*) also accounts for about 0.5% of the value of the crop

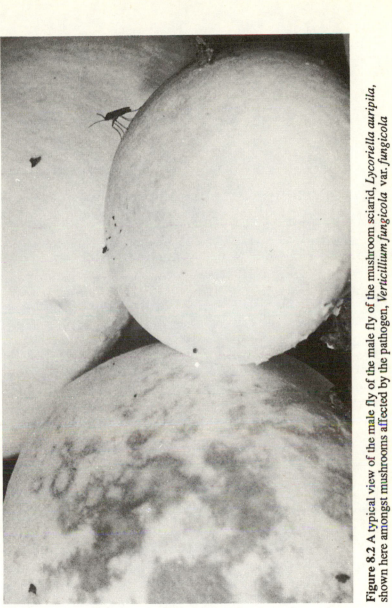

**Figure 8.2** A typical view of the male fly of the mushroom sciarid, *Lycoriella auripila*, shown here amongst mushrooms affected by the pathogen, *Verticillium fungicola* var. *fungicola*

**Figure 8.3** Adult (female) of the mushroom sciarid, *Lycoriella auripila*

**Figure 8.4** The abdomen of a male sciarid fly, *Lycoriella auripila*, showing prominent claspers at its tip

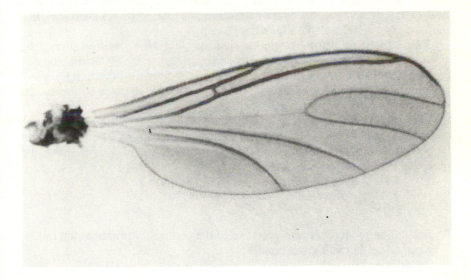

**Figure 8.5** Typical wing venation, *Lycoriella auripila*

(on an 'average' mushroom farm), this figure represents the economic threshold for this pest.

CONTROL

Access of sciarid adults to peak-heat and spawn-running rooms must be prevented with the use of screening on ventilation ducts (16 mesh/cm, 40 mesh/inch) and efficient sealing around doors. The number of flies that get past these physical barriers should be monitored with the use of sticky traps, which are examined daily.

Experiments have shown that an antagonism exists between sciarids and mycelium, such that large volumes of mycelium inhibit larval development, and a compost well-colonized by mycelium does not favour oviposition by females. Thus, a vigorous spawn-run throughout the compost will, by deterring sciarid oviposition and development, exert a degree of cultural control. Conversely, a poor compost, through which mycelium has difficulty in growing, will provide an attractive substrate for sciarid development.

Where resistance to organophosphorous chemicals is not a problem, diazinon or chlorfenvinphos can be mixed into the compost at spawning.

Newly spawned crops should be protected for the first few days by aerosols, smokes or fogs of pyrethrins, gamma-HCH or dichlorvos* to control ingressing flies (see Table 8.1). Casing material should be treated with diflubenzuron to kill larvae. To gain maximum effect, it should be incorporated during the preparation of the casing. Diflubenzuron can also be used as a post-casing drench but, to ensure maximum penetration of the chemical, this must be done as soon as possible after casing.

Flies emerging before cropping (usually just after casing) should be controlled by aerosols, smokes or fogs of pyrethrins, gamma-HCH, dichlorvos* or pirimiphos-methyl*. For flies that emerge during the cropping period, pyrethrins are preferable as there is less risk of toxic damage to the mushrooms and less nuisance to pickers.

If sciarids develop on an untreated crop, then OP-susceptible populations of larvae can be killed with a surface drench of malathion. The beds should be picked hard before treatment, and a harvest interval of 4 days should be observed to ensure freedom from taint.

ACTION POINTS

● Ensure the production of a good, selective compost through which mushroom mycelium will grow rapidly;

---

* These chemicals will not control OP-resistant populations.

- Use sticky traps to monitor flies in the spawn-running room;
- Use effective screening of ventilation ducts in the peak-heat and spawn-running rooms;
- Use knock-down sprays during the first week of spawn-run and the first week after casing;
- Incorporate diflubenzuron in the casing;
- Use pyrethrins during cropping;
- At the end of cropping, ensure that an effective cook-out temperature is reached throughout both the growing medium and the trays or shelves;
- Make sure that all the spent compost is removed from the farm.

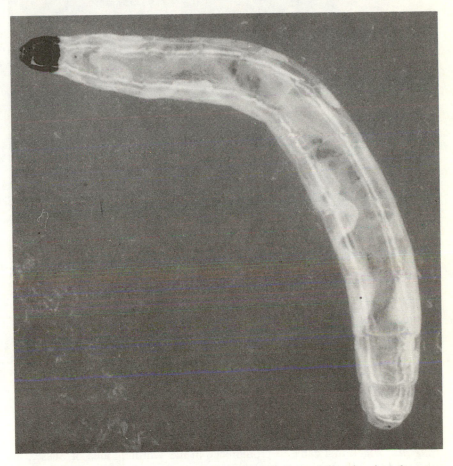

**Figure 8.6** Larva of the mushroom sciarid, *Lycoriella auripila*, showing characteristic black head

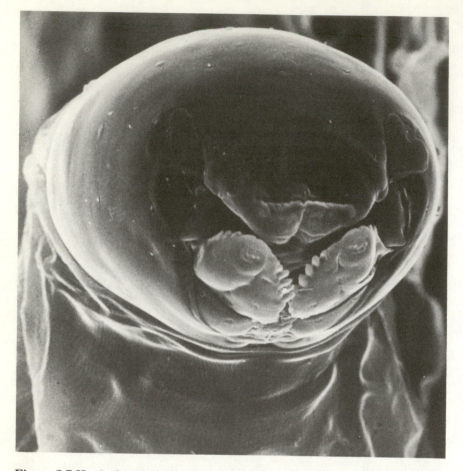

**Figure 8.7** Head of a sciarid larva *(Lycoriella auripila)* showing large mouthparts (scanning electron micrograph)

## Phorids: *Megaselia halterata*

Of the 99 species of *Megaselia* that are known to exist in the United Kingdom, only two are pests in the mushroom industry. The most important species is *M. halterata*, which first came to prominence in the Worthing area of Sussex in 1953, and for 28 years it was the dominant mushroom pest.

SYMPTOMS

*M. halterata* larvae feed solely on mushroom mycelium and, consequently, are able to cause a reduction in yield. However, as the larvae do not burrow into mushrooms and develop mostly in the compost, they are seldom seen. Rather, it is the flies which cause the most obvious symptoms. The flies are normally most numerous in summer and late autumn. They are very active near lights and, therefore, can be a considerable nuisance to pickers. They are also vectors of *Verticillium* (*see* page 56), and a density of only 75 flies/m$^2$ (7/ft$^2$) is capable of spreading the pathogen and initiating the disease. This density is more than 130 times less than the number of adults required to cause economic crop damage (by their subsequent larvae). Thus, by acting as a vector for this disease, flies pose a greater threat to the mushroom crop than do the larvae.

DESCRIPTION OF PEST

Mushroom phorids are small (2–3 mm), hump-backed flies with inconspicuous antennae (*Figure 8.8* and *Frontispiece*). They resemble diminutive house flies, are brown-black in colour, and generally stouter in appearance than sciarids. There is no obvious difference between the male and female flies, although, on closer examination, the male abdomen ends in a black capsule, while the female abdomen is generally paler and its apex is pointed. The wing venation is also a characteristic identification feature (*Figure 8.10*). The flies congregate on the surface of boxes or shelves — often at the top corners of a stack of trays, and especially near lights or doors. When seen on compost or other surfaces, they scuttle about with a characteristic rapid, jerky run.

The larvae are creamy-white, legless maggots (1–6 mm long) with a pointed head end which is not black, and a blunt rear end (*Figure 8.9*) (cf. sciarid larva, *Figure 8.6*). Two-thirds of their immature life is spent as an immobile, non-feeding pupa (*Figure 8.11*), which is 2–3 mm long and varies in colour from creamy-white to brown as the fly inside develops. They can sometimes be seen in the compost, especially at the sides and corners of the boxes.

EPIDEMIOLOGY

*M. halterata* is unable to fly when the air temperature is below (54°F) and therefore, in general, populations from the wild are unlikely to invade a mushroom farm between November and March. Short, warm spells during this period can, however, bring about isolated infestations. This highlights the importance of continually monitoring the level of flies invading a farm. Once the air temperature exceeds (63°F), temperature is no longer a limiting factor.

**Figure 8.8** Adult of the mushroom phorid, *Megaselia halterata*

**Figure 8.9** Larva of the mushroom phorid, *Megaselia halterata*

Windy weather curtails flight — but not crawling — while rain needs to be quite heavy to prevent activity as, in still conditions, flight may occur in drizzle, or even light rain. Phorids do not fly after twilight, no matter how high the temperature.

Once they have mated, female *M. halterata* are attracted to the smell of growing mycelium and, therefore, pose a threat as soon as the crop is spawned. They fly upwind to the source of the aroma and, once inside a spawn-running room, they locate the compost and lay up to 50 eggs in close proximity to the growing tips of the mushroom mycelium. The compost is attractive for about 3 weeks after spawning, with the maximum attraction occurring in the second week. After the crop is cased, fresh mycelial growth occurs, again attracting oviposition by females.

The rate of development, from egg to adult, varies considerably with the compost temperature. At a temperature of 24°C (75°F) (spawn-running), it takes only 15 days for the insect to develop from egg to adult. During cropping, compost temperatures are generally cooler (15–20°C, 60–70°F), resulting in increased generation times (50–24 days respectively as the temperature increases). This range of temperatures explains why phorids can emerge at any period during cropping, although the majority would probably emerge 3–4 weeks after cropping starts.

The economic threshold for larval density has not been determined, although early work indicated that 100 larvae/30 g compost can seriously affect mycelial growth in compost. Some more recent work done in the U.S.A. on the same species has indicated that the economic threshold for fly density at spawning was an infestation of between 10,000 and 12,000 flies/m². If this threshold is translated into a potential larval density, then *M. halterata* larvae would need to be present at a density 58 times that of sciarid larvae to cause the same degree of crop loss.

CONTROL

It is essential for the spawn-running and pre-cropping periods to be protected whenever conditions are suitable for flight. Dichlorvos is a useful chemical as it has both contact and volatile action on flies. One application should last 6 hours, so it follows that, for full protection against ingressing adults, twice-daily applications during daylight hours (e.g. 8–9 a.m. and 2–3 p.m.) are required *(see Table 8.1)*. Aerosols and/or smokes of pyrethrins or gamma-HCH can also be used, although their effects are very short-lived. This is a great disadvantage if flies are continually entering a building during suitable 'flight' weather.

If fresh air is used for cooling purposes, it is unlikely that any sort of aerial spraying system will be effective.

In any attempt to exclude flies by physical means from the spawn-running and case-running rooms, careful attention should be given to screening ventilation ducts, especially the air outlets, as it is these that are the source of the aroma that attracts the flies. Because the females are able to squeeze through

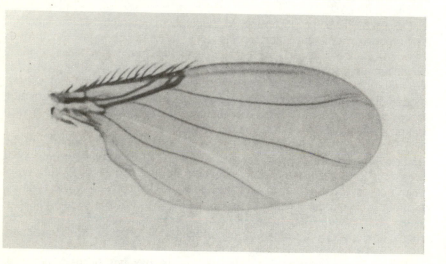

**Figure 8.10** Typical wing venation of *Megaselia halterata*

very small cracks and crevices, the screening should have a mesh of 16 apertures/cm (40/inch) or less. The sealing around screens, doors and cracks in walls should be exceptionally thorough.

To kill larvae in the compost, diazinon or chlorfenvinphos *(see Table 8.1)* must be incorporated at spawning. Whichever chemical is used, it is essential that a thorough mixing and even distribution of the insecticide is achieved throughout the whole of the compost, as the degree to which phorids are controlled will reflect merely the efficiency of the mixing. Care should be taken to keep the insecticide and spawn apart as far as possible during the spawning process because, if they come into intimate contact, the insecticide will have an adverse effect on subsequent mycelial growth and cropping. They should not, therefore, be mixed in the hope of simplifying application. Irrespective of the formulation of insecticide used, it should be applied to the compost before the mushroom spawn is added. This will help to avoid 'coating' the spawn with the insecticide.

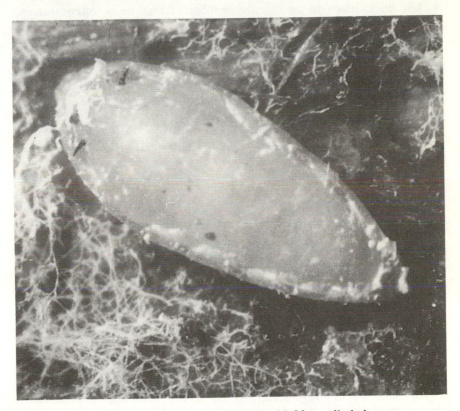

**Figure 8.11** Pupa of the mushroom phorid, *Megaselia halterata*

Treatment of the compost protects only the spawn-run. Failure to protect the crop after it has been cased can result in an infestation of the casing and subsequent damage to the mycelium at the compost/casing interface. Such damage may be prevented by aerial treatments to kill ingressing flies (as described above).

Flies emerging during cropping can be controlled by fogs, smokes or aerosols of pyrethroid-related compounds, although routine use of such compounds can result in resistance. It cannot be over emphasized that the control of flies during cropping, apart from reducing the risk of re-infestation and disease spread, is merely a cosmetic operation if used alone. Such measures must, therefore, be part of a properly conceived and executed control programme.

ACTION POINTS

- Use sticky traps to monitor the number of flies in the spawn-running room;
- Use effective screening of ventilation ducts in the spawn-running room;
- Thoroughly seal gaps or cracks in the doors and walls of the spawn-running room;
- Protect crops for 3 weeks after spawning with applications of dichlorvos;
- Incorporate diazinon into the compost at spawning;
- Use pyrethrins during cropping to kill flies;
- At the end of cropping, ensure that an effective cook-out temperature is reached throughout both the growing medium and the trays or shelves;
- Make sure that all the spent compost is removed from the farm.

## Phorids: *Megaselia nigra*

SYMPTOMS

*M. nigra* larvae feed on and in developing mushrooms. After the eggs have hatched, the larvae tunnel into the mushroom tissues and develop rapidly, taking only 5 days at 18°C (64°F) to develop into a pupa. The fly emerges from the pupa after a further 5 days. The cap and the stipe can be tunnelled by dozens of larvae *(Figure 8.12)* and, in extreme cases, the sporophore can rapidly deteriorate, due to bacterial decay spreading from the sites of larval attack.

This sort of damage should not be confused with the damage which occurs with sciarid larvae *(see Figure 8.1)*. *M. nigra* damage normally occurs from the top of the mushroom downwards, whereas sciarid larval damage occurs from the bottom of the stipe upwards, only rarely burrowing within the cap.

**Figure 8.12** Typical damage caused by the larvae of *Megaselia nigra*

DESCRIPTION OF PEST

The adults of *M. nigra* are slightly larger and darker than those of *M. halterata*, but otherwise have the same general characteristic appearance. The larvae of both species are similar, except that those of *M. nigra* are generally longer and possess a pair of distinct, black 'mouth-hooks' in their pointed heads with which they are capable of burrowing within mushroom tissue.

Adult females usually lay their eggs on the gills of developing mushrooms, although they can also be laid on the casing surface or immediately next to the developing pin-heads.

EPIDEMIOLOGY

*M. nigra* can readily be found in wild mushrooms, where tunnelling is common, and it has always been a pest of commercial mushroom growing. Light traps have shown that adults fly from June to December, and that they are likely to be troublesome in late summer. *M. nigra* lays its eggs only where natural daylight falls on to a mushroom bed. In most purpose-built mushroom houses, therefore, it is unlikely to be a problem. However, where doors are opened for cooling purposes in the summer, damage may occur.

Table 8.1 Insecticides for mushroom pest control

| Common name | Trade name | Method and rate of use | Pests controlled |
| --- | --- | --- | --- |
| Diazinon (Ciba-Geigy) | Basudin 5 FG | Mix with compost at spawning 200 g/tonne (7 oz/ton) | Phorids* |
| | Basudin 40 WP | 1 kg/tonne (36 oz/ton) | Sciarids* |
| | | 25 g/tonne (1 oz/ton) | Phorids* |
| (Dow) | Diazinon Liquid | 125 g/tonne (4.5 oz/ton) | Sciarids*† and cecids* |
| | | 56 ml/tonne (2 fl oz/ton) | Phorids* |
| Dichlorvos (Ciba-Geigy) | Darmycel Dichlorvos | Aerial spray in spawn-running rooms 30 ml in 300 ml water/140 m³ (1 fl oz in 0.5 pint water/5000 ft³) | Phorids and Sciarids† |
| Diflubenzuron (ICI) | Dimilin 25 WP | Mix with casing 120 g/tonne (4.5 oz/ton) Drench to casing** | Sciarids |
| | Dimilin 25 WP | 4g in 2.5 l water/m² (13.5 oz in 50 gal water/1000 ft²) | Sciarids |
| Chlorfenvinphos (Ciba-Geigy) | Sapecron 240 EC | Mix with compost at spawning 208 ml/tonne (7.5 oz/ton) | Sciarids*† and Phorids* |
| | Sapecron 10 FG | 500 g/tonne (18 oz/ton) | Sciarids*† and Phorids* |

| | | | |
|---|---|---|---|
| Malathion (Farm Protection) | Malathion 60 | Drench to casing 330 ml in 200 l water/100 m² (10.5 fl oz in 40 gal water/1000 ft²) | Sciarids† |
| Gamma-HCH (Octavius-Hunt) | Fumite Lindane Pellets | Aerial smoke during cropping Pellet size 3, treats 84 m³ (3000 ft³) | Sciarids and Phorids |
| Resmethrin/pyrethrins (Mitchell-cotts) | Pynosect 30 | Aerial spray during cropping 30 ml/100 m² (1 fl oz/1000 ft²) | Sciarids and Phorids |
| Resmethrin (PBI) | Turbair resmethrin extra | Aerial spray during cropping 30 ml/100 m² (1 fl oz/1000 ft²) | Sciarids and Phorids |
| Permethrin (PBI) (Octavius-Hunt) | Turbair permethrin Fumite permethrin smokes | Aerial spray during cropping 30 ml/100 m² (1 fl oz/1000 ft²) Canister size 4000, treats 112 m³ (4000 ft³) | Sciarids and Phorids Sciarids and Phorids |
| Pirimiphos-methyl (ICI) | Actellifog | Aerial fog during cropping 40 ml/100 m³ (4 fl oz/10 000 ft³) | Sciarids† and Phorids |

† Sciarid populations that are resistant to organophosphorous insecticides will not be controlled with these chemicals

* Manufacturer's label recommendations differ from those quoted above

** These figures are based on a prepared casing weight of 1 tonne/30 m³ (3 tons/1000 ft²)

CONTROL

Because *M. nigra* lays its eggs on, or near to, developing mushrooms and there would be a risk, therefore, of toxic residues, no pesticide treatments are appropriate. To prevent egg-laying, ensure that the mushroom beds are not exposed to natural daylight.

ACTION POINT

● Black-out any openings in the building structure which allow natural daylight to fall on the mushroom beds.

## Cecids

Six species of cecid have been recorded on mushrooms, but only three are common: *Heteropeza pygmaea, Mycophila speyeri* and *M. barnesi.* Of these three, *H. pygmaea* occurs most frequently.

SYMPTOMS

The white or orange larvae are normally first noticed when, after watering, they swarm on to mushrooms, where they feed on the outside of the stipes or at the junction of the stipe and gills *(Figure 8.13).* With *H. pygmaea*, the loss in marketable yield, due to spoilage, is attributable not only to the presence of the larvae on the mushrooms, but also to bacteria which are present on their skin. These cause brown, discoloured stripes on the mushroom stipe and gills *(Figure 8.14).* The delicate gill tissue can then break down to produce tiny pustules of black fluid.

DESCRIPTION OF PEST

Cecids are rarely identified from the fly stage because they are minute and seldom seen (Figure 8.15). Usually, they are identified from the larvae, which are legless maggots and either white *(H. pygmaea)* or orange (*Mycophila* spp.) *(Colour Plate 14).* They have no discernible head, but there are two 'eye-spots' at the head end which, together, give the appearance of an 'X'.

Reproduction in these mushroom cecids is unusual in that it is normally achieved by a process called paedogenesis. Whereas in most insects a larva is destined to become a pupa and then an adult fly, each cecid larva becomes a 'mother-larva' which will give birth to 12–20 'daughter' larvae within a week of its own birth, without any adult cecid being present (*Colour Plate 15* and *Figure 8.16*). Because several stages of a normal reproductive life cycle are bypassed, this method of reproduction leads to a very rapid multiplication. Consequently, enormous populations — as high as 18,000 per handful of casing — can develop.

EPIDEMIOLOGY

As mushroom cecids can be readily found in decaying wood and rotten vegetation, it is possible that the flies from such natural habitats may give rise to an infestation on a farm. However, initial, very small infestations probably arise from infested peat. After developing, unnoticed, through a number of generations, subsequent spread of the larvae about a farm will give rise to further, more serious outbreaks. Spread of the small, sticky larvae occurs on inadequately sterilized tray or bed timbers, and on the hands, tools, equipment, shoes and clothes of workers.

The commonest species is *H. pygmaea* and, when present in large numbers, can swarm out of the beds and fall on to the floor, where they

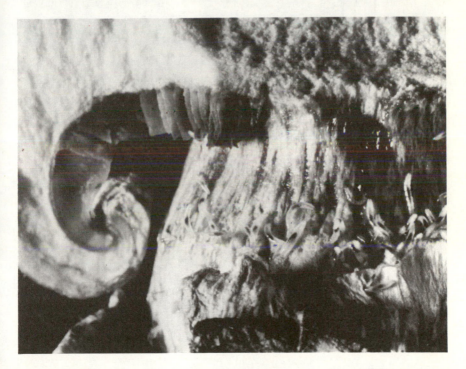

**Figure 8.13** Larvae of the mushroom cecid, *Heteropeza pygmaea*, feeding on a mushroom

**Figure 8.14** Damage to mushroom gills caused by the mushroom cecid, *Heteropeza pygmaea*

congregate, if the floor is dry, in writhing heaps *(Figure 8.17)*. In addition to the loss in marketable yield caused by their swarming on the mushrooms — mostly in the latter half of the crop — they can also cause a loss in total yield, due to the action of the larvae feeding on the mycelium.

The two *Mycophila* species can also cause both spoilage and depression of yield. Because they breed more rapidly than *H. pygmaea* and are very conspicuous, being orange in colour, they can cause up to 50% spoilage of mushrooms. *M. speyeri* generally swarms in the first and second flushes and, as the crop progresses, they are liable to disappear. *M. barnesi* generally swarms in the middle to later flushes.

**Figure 8.15** Female fly of the mushroom cecid, *Heteropeza pygmaea*. (Reproduced by kind permission of Dr I. J. Wyatt)

**Figure 8.16** Hemipupa or 'mother-larva' of the mushroom cecid, *Heteropeza pygmaea*. The 'daughter-larvae' can be seen inside the hemipupa. (Reproduced by kind permission of Dr I. J. Wyatt)

CONTROL

Larvae are easily spread about the farm, so it is vital to isolate infested houses by the use of 'Sudol' foot dips *(see* page 32) and to disinfect tools, etc., with the same chemical. The use of separate protective overalls in every house is also an advisable precaution. *H. pygmaea* is unusual in that, when the food supply starts to become exhausted, it can produce especially adapted mother larvae which have hard brown skins. Within this protective 'shell', one or two daughter larvae are able to survive for several months without food. It is, therefore, very important to ensure that spent compost is efficiently cooked-out and removed from the farm.

Efficient peak-heating and cooking-out should kill all larvae, as *H. pygmaea* is killed at a temperature of 45°C (113°F), and *Mycophila* spp. at slightly lower temperatures. If cooking-out is not possible, then the finished crop and houses should be fumigated with methyl bromide. Both treatments must penetrate the wood of shelves or trays to be effective.

Dipping empty trays and bed boards in 2% sodium pentachlorophenoxide *(see Table 3.2,* page 34) and treatment with live steam will kill superficial mycelium on which cecids might survive and re-infest compost at spawning. However, these treatments will not kill larvae or fungi deeper within the wood.

**Figure 8.17** Clumps of mushroom cecid larvae *(Heteropeza pygmaea)* which have swarmed out of a mushroom bed on to the floor of the house

Empty houses should be thoroughly washed down with a disinfectant (*see* page 30). Casing materials should be hygienically stored before use; if these are suspected to be harbouring pests, then they should be fumigated with methyl bromide, or pasteurized with steam.

*H. pygmaea* larvae cannot be killed with insecticides, but the incorporation of diazinon in the compost will reduce larval breeding and, therefore, contamination (*see Table 8.1*). This treatment will kill the larvae of *Mycophila* spp.

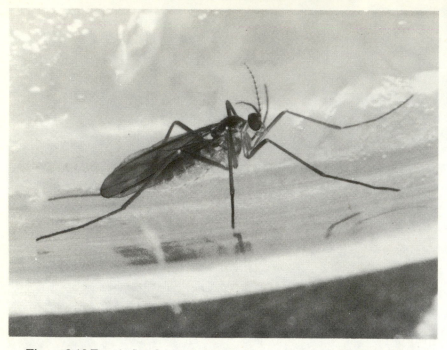

**Figure 8.18** Female fly of the non-paedogenetic cecid, *Lestremia leucophaea*

ACTION POINTS

- Observe strict hygiene throughout the farm;
- Make sure that the casing ingredients are stored and mixed in a clean area;
- Isolate infested houses with Sudol foot dips;
- Incorporate diazinon in the compost at spawning;
- Dip wooden trays in 2% sodium pentachlorophenoxide;
- At the end of cropping, ensure that an effective cook-out temperature is reached throughout both the growing medium and the trays or shelves;
- Make sure that all the spent compost is removed from the farm.

## Cecids (non-paedogenetic)

Apart from the main paedogenetic cecids, two species of non-paedogenetic cecid, *Lestremia cinerea* and *L. leucophaea,* are occasional pests of mushrooms.

SYMPTOMS

The orange or pink larvae can be found feeding on the developing mushrooms. *L. leucophaea* is found mostly at the base of the stipe.

DESCRIPTION OF PEST

The body of the adult female fly is 3–4 mm long, depending on the species, and its swollen abdomen, which is packed with eggs, is a dull orange colour. As the legs are very long and slender and, in *L. leucophaea,* have a span of about 12 mm *(Figure 8.18)*, both species bear a marked resemblance to mosquitoes. The larvae of both species attain a length of about 4 mm when fully grown. *L. cinerea* larvae are dark orange in colour, whereas *L. leucophaea* larvae are a delicate salmon-pink.

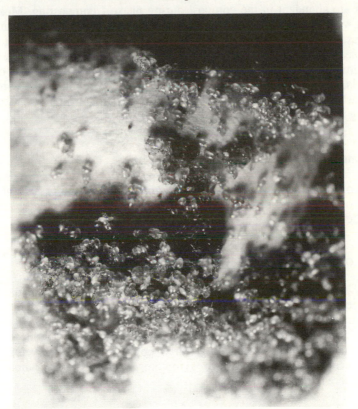

**Figure 8.19** A clump of tarsonemid mites *(Tarsonemus myceliophagus)*, highly magnified, on mushroom tissue

Both species are common in the wild and there is evidence, especially with *L. cinerea*, that the females require natural daylight to lay their eggs (cf. *Megaselia nigra* page 127).

No specific control measures are appropriate, although adult flies will be killed by the same chemicals used for phorid and sciarid control.

## Mushroom mites

Mites are abundant in the compost during phase I composting. Most of the easily visible species are predatory, feeding mainly on nematodes and other species of mite which are present at this time. An efficient peak-heat will kill all mites. Of the mites to be found in a crop from spawning onwards, most do not directly affect the crop: some are predatory; some feed solely on weed moulds, and some feed on bacteria. The last, when found in the compost or on mushrooms, are not necessarily initiating decay, but often are merely following it.

## Tarsonemid mites

The most important mite pest of mushrooms is *Tarsonemus myceliophagus*. It is the only mushroom mite where clear evidence of primary pest status has been demonstrated.

The mites cause a reddish-brown discoloration and rounding of the base of affected mushroom stipes — sometimes severing the basal attachments of the mushroom entirely *(Colour Plate 16)*. As the mites reproduce slowly, damage normally occurs after about 4 weeks' cropping. However, where an infestation occurs soon after spawning, large populations may develop, with a subsequent increase in damage symptoms, such that the whole of the sporophore may be discoloured.

DESCRIPTION OF PEST

*T. myceliophagus* is a pale brown, shiny mite *(Figure 8.19)*, which is so minute (0.18 mm) that it is virtually invisible to the naked eye. Compared with most mites and insects, it has a fairly slow rate of increase, producing (on average) one egg/day over a period of 3 weeks. The mites feed on the hyphae both of mushrooms and of various weed moulds *(Chaetomium, Trichoderma and Penicillium)*, developing more rapidly on the last.

CONTROL

To kill mites at the end of cropping, a normal cook-out should be ample because a 20-minute exposure to 49°C (120°F) is lethal to them. However, live mites have been found in crops after such a treatment: therefore, niches must occur which harbour the mites and where the temperature-time product does not exceed the lethal value. A thorough and efficient cook-out with steam or fumigation with methyl bromide is, therefore, essential. Care and general hygiene of the houses, especially in the clearance of debris, is also important. Dicofol is very effective when used as a disinfectant for washing all surfaces such as floors, walls, tools, filling and spawning lines.

ACTION POINTS

- Observe strict hygiene throughout the farm;
- Ensure that all machinery and rooms involved with spawning and the spawn-running process are thoroughly cleaned;
- At the end of cropping, ensure that an effective cook-out temperature is reached throughout both the growing medium and the trays or shelves;
- Make sure that all the spent compost is removed from the farm.

## Red pepper mites

The most common mites seen during mushroom cropping are the red pepper mites. These are species of *Pygmephorus* and are not normally regarded as primary pests. Three species have been recorded on mushrooms in Britain.

SYMPTOMS

These mites often swarm in vast numbers on the surface of the casing and the mushrooms, giving them a reddish-brown colour *(Colour Plate 17)*, and it is from such behaviour that they get their common name. They feed only on weed moulds (particularly *Trichoderma* spp.) and not on mushroom mycelium. The mites' presence indicates a poor compost in which weed moulds are abundant and, although not regarded as a primary pest, they have been implicated in the spread of *Trichoderma harzianum* (*see* page 67).

DESCRIPTION OF PEST

These mites are tiny (0.25 mm), yellowish-brown in colour, with a central whitish internal band, and have a flattened, wedge-shaped appearance (*Figure 8.20*). They are capable of rapid rates of increase as females can lay up to 160 eggs over a period of 5 days.

**Figure 8.20** A clump of red pepper mites (*Pygmephorus* spp.), on mushroom tissue, showing internal whitish band

CONTROL

Efficient composting and peak-heating of fresh manure, to produce a medium selective for mushroom mycelium, is the most important way of avoiding trouble with red pepper mites. Because of their reported ability to spread *T. harzianum* (*see* page 67), it may be considered desirable to treat isolated areas of red pepper mites. Although both salt and flame guns have been used with some success, it should be emphasized that surface treatments will not correct the basic underlying compost problem.

ACTION POINTS

- Observe strict hygiene throughout the farm;
- Make sure that composting and peak-heating are efficient.

## Saprophagous mites

Included in this group of mites are *Tyrophagus* spp., *Caloglyphus* spp., and *Histiostoma* sp., all of which are saprophagous species of mite that occasionally can be found on mushrooms.

### SYMPTOMS

In the past, some of these mites (especially the *Tyrophagus* spp.) have been associated with small pits in the caps and stipes of mushrooms, which then often suffer from bacterial decomposition (*see* page 81). It is not clear, however, whether the mites are causing the damage, or are merely exacerbating an existing bacterial attack. These mites may also feed on the mycelium and have been shown to have a limited capability to spread the spores of *Verticillium*.

### DESCRIPTION OF PEST

These are soft, translucent, white mites whose bodies carry long hairs (except *Histiostoma* sp.). They are larger (0.3–0.5 mm) and much slower in their movements than tarsonemid and red pepper mites. They are saprophagous and normally are only found in foul compost into which mushroom mycelium cannot grow. Consequently, they are often found in association with saprophagous nematodes (*see* page 145).

### EPIDEMIOLOGY

*Tyrophagus* spp. are usually introduced into the compost by flies on whose bodies the migratory stages cling by means of suckers. This migratory stage is normally produced when the mites become overcrowded.

### CONTROL

Where efficient composting and peak-heating are carried out, saprophagous mites are unlikely to be a nuisance. However, fungal and bacterial contaminants are commonly found at the end of the cropping period and, as these mites breed readily on such substrates, an efficient cook-out and safe disposal of the spent compost is essential. Good general hygiene — especially in the clearance and elimination of all organic debris — is also important.

ACTION POINTS

- Observe strict hygiene throughout the farm;
- Make sure that composting and peak-heating are efficient;
- Control flies;
- At the end of cropping, ensure that an effective cook-out temperature is reached throughout both the growing medium and the trays or shelves;
- Make sure that all the spent compost is removed from the farm.

## Predatory mites

There are three species of predatory mesostigmatid mites (often called gamasid mites) which are frequently encountered in mushroom houses. They are *Parasitus fimetorum*, *Digamasellus fallax* and *Arctoseius* cetratus and, as they feed on various stages of most mushroom pests, they are beneficial and should not be regarded as pests. They can, however, be regarded as indicator species because, if they are present in large numbers, their food source (the pests) is also abundant.

SYMPTOMS

Predatory mites cause no symptoms on the mushroom crop but, as they can easily be seen running over the surface of the casing, mushrooms and trays in search of their prey, they can be a source of irritation to pickers.

DESCRIPTION OF MITES

These mites are pale orange to dark red, and are most easily distinguished from pest species by their larger size and/or their greater speed of movement. They can range in length from 0.35 mm *(A. cetratus)* to 1.1 mm *(P. fimetorum)*. These two mites are polyphagous and will accept a wide range of prey, including insect eggs and larvae, mites and nematodes *(Figure 8.21)*.

EPIDEMIOLOGY

*P. fimetorum*, *D. fallax* and A. *cetratus* can all be introduced into a mushroom crop by flies, especially sciarids, as they are able to attach themselves to the abdomens of their more mobile hosts *(Figure 8.22)*.

**1.** Wet bubble disease (*Mycogone perniciosa*) with typical amber droplets on the distorted mushrooms

**2.** Dry bubble disease (*Verticillium fungicola* var. *fungicola*) distortion and cap spotting

**3.** Dry bubble disease (*Verticillium fungicola* var. *fungicola*) severe distortion mushroom

**4.** Cap spotting caused by (*Verticillium fungicola* var. *aleophilum*)

**5.** Decay and discoloration of mushrooms resulting from Cobweb disease *Cladobotryum dendroides*

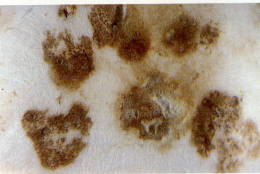

**6.** Cap spotting caused by *Cladobotryum dendroides* (*Dactylium dendroides*)

**7.** Cap spotting caused by *Trichoderma konongii*

**8.** Compost colonized by *Trichoderma harzianum*. Note the green mould of the fungus and the red pepper mites in the corner

**9.** Cap spotting caused by *Aphanocladium album*

**10.** Brown blotch (*Pseudomonas tolaasii*)

**11.** Virus disease caused by mushroom virus 4

**12.** Virus disease showing typical stalk elongation and small caps

**13.** Olive-green mould  (*Chaetomium olivaceum*) and mushroom mycelium

**14.** Three larvae of the mushroom cecid, *Mycophila speyeri*, showing their range in size and distinctive orange colour. (Reproduced by kind permission of Dr I. J. Wyatt)

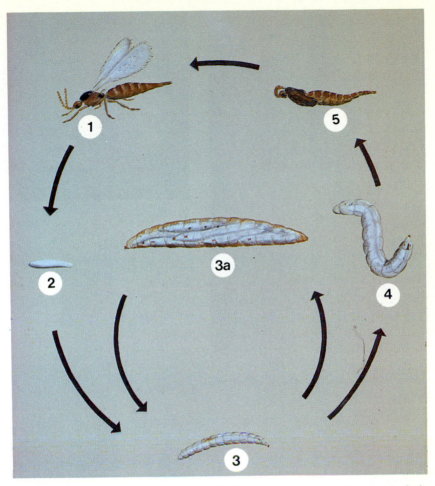

**15.** The life cycle of the mushroom cecid, *Heteropeza pygmaea*, showing both the paedogenetic and sexual phase of reproduction: (1) Female fly; (2) Egg; (3) Larva; (3a) Paedogenetic 'mother' larva; (4) Pupa larva; (5) Pupa. (Reproduced by kind permission of Dr I. J. Wyatt)

**18.** Rosecomb

**16.** Typical damage caused by the mite *Tarsonemus myceliophagus*. The base of the stipe is discoloured with only partial attachment to the casing

**17.** Mushrooms covered with jostling swarms of the red pepper mite, *Pygmephorus* spp.

## CONTROL

As these mites are beneficial, no measures should be taken to control them. Rather, the crop should be examined for their food source and, on these findings, appropriate measures should be taken against the actual pest that is present.

### ACTION POINT

● Identify the pest or pests on which the predatory mites are feeding and take appropriate control measures.

## Nematodes

Many different species of nematode can be found associated with decaying organic matter. Thus, the compost and casing material are favourable environments for those nematodes that feed on decaying matter (saprophagous nematodes) and for those that are specifically fungal feeders (mycophagous nematodes).

### DESCRIPTION OF PEST

They are small, colourless worms up to 1 mm long *(Figure 8.23)*, that swim in surface films of water in the compost and casing and are, therefore, normally visible only with the aid of a microscope. In favourable conditions, they can

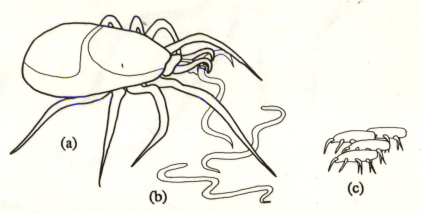

**Figure 8.21** Representation of: (a) Predatory mite; (b) Nematodes; (c) Red pepper mites; showing the relative sizes of each

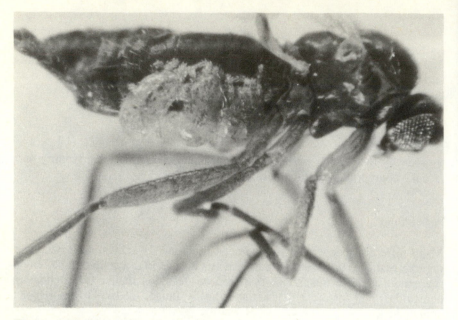

**Figure 8.22** Female fly of the mushroom sciarid, *Lycoriella auripila*, showing predatory mites attached to its abdomen

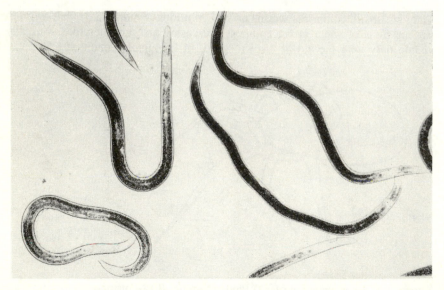

**Figure 8.23** Rhabditid (saprophagous) nematodes

increase very rapidly (50–100–fold per week), so that their potential importance to the crop is considerable, depending upon which species is present.

## Saprophagous nematodes

The saprophagous nematodes, mostly rhabditids, form the largest group of nematodes to be found in mushroom beds. They are not normally regarded as primary pests because they feed on decaying plant material. Their presence within a crop, therefore, is chiefly an indication of poor growing conditions.

SYMPTOMS

An inefficient peak-heat will produce conditions which are unfavourable for the growth of mushroom mycelium and may also allow nematodes to survive. These can combine to produce an abnormal nematode population in the compost at spawning, as well as poor subsequent cropping. They can swarm in vast numbers on the surface of the casing, where they can be seen glistening in the light *(Figure 8.24)*. Bad harvesting practices, such as leaving cut stipes or chogs on the beds, can also encourage the development of these nematodes as they will feed on the decaying mushroom tissue *(Figure 8.25)*.

Recent commercial experience has shown that, if 'off-white' strains of mushrooms are given a short spawn-run, nematodes introduced in a contaminated casing medium can cause a direct loss of yield. There is also some evidence that, if the compost is heavily infested with nematodes at spawning, other strains may be similarly affected.

## Mycophagous nematodes

*Ditylenchus myceliophagus* and *Aphelenchoides composticola*, two species of mycophagous nematode, are primary pests of the mushroom because they feed exclusively on fungi and, therefore, can destroy mushroom mycelial growth. Under normal hygienic conditions and short cropping periods, however, they do not often cause serious trouble.

SYMPTOMS

If contamination occurs early, the numbers of nematodes can reach a level sufficient to destroy the mycelium. Subsequent bacterial action causes infested compost to become dark and sodden, often in defined patches, with a distinctive pungent smell. These patches can increase in size from the second flush onwards.

**Figure 8.24** A swarm of rhabditid (saprophagous) nematodes on the surface of a mushroom bed

The extent of the loss in yield is dependent on the time and level of the original infestation. An infestation at spawning can cause a failure of the spawn-run and, thus, can render growing completely uneconomic. A later infestation towards the end of cropping would, however, cause only a slight loss in yield, and would probably go unnoticed by the grower.

*A. composticola* feeds on many fungi, as well as mushrooms. It breeds very rapidly and, in heavy infestations, the nematodes tend to adhere together in whitish clumps.

CONTROL

Both nematodes and cecids occasionally occur in peat when it is cut. Therefore, if the peat is suspected as being the source of infestation, all casing peat that comes on to the farm should be checked. Where insects or nematodes are present, then dry peat should be fumigated with methyl bromide or pasteurized by steam. It should then be protected from further pest infestation until it is used for casing.

Once an attack by mycophagous nematodes has occurred, little can be done, apart from destroying the compost from affected areas, plus some of the surrounding, apparently healthy, compost. Cooking-out at the end of the crop is the best method of controlling nematodes. However, they are extremely difficult to eradicate and can be carried over from one crop to another, lodged in cracks in the bed boards and trays: a uniform temperature of 55–60°C (131–140°F) must be achieved, therefore, throughout the compost, trays and bed boards. If effective cooking-out is not possible, the compost should be fumigated with methyl bromide.

*D. myceliophagus* can withstand drying for up to 3 years, and this makes the safe disposal of all spent compost very important as dry, infested debris is a potential source of trouble.

ACTION POINTS

- Observe strict hygiene throughout the farm;
- Ensure that the temperatures during peak-heat are satisfactory;
- Make sure the casing ingredients are stored and mixed in a clean area;
- At the end of cropping, ensure that an effective cook-out temperature is reached throughout both the growing medium and the trays or shelves;
- Make sure that all the spent compost is removed from the farm.

**Figure 8.25** Rhabditid (saprophagous) nematodes feeding on a decaying mushroom stalk left on the casing surface

**Figure 8.26** Typical damage to a mature sporophore by springtails

## Minor pests

There are several other organisms which are pests of mushrooms only occasionally. Their presence can sometimes be associated with bad growing conditions or unconventional growing systems, as most of them are either saprophagous or ground-dwelling in nature.


## Collembola (springtails)

If mushroom beds are made up on the floor, springtails may be a pest. They favour damp conditions with an abundance of decaying vegetable matter: thus, catch-crops grown in glasshouses were especially vulnerable. In such conditions, enormous numbers may develop, to such an extent that they may cover the surface of the bed.

The commonest species is *Archorutes armatus*, which is slate-blue in colour. It can feed on both mycelium and sporophores, resulting in minute open pits in the stem and cap *(Figure 8.26)*. From these pits, dry branched tunnels can be formed.

**Figure 8.27** Adult fly of the sphaerocerid, *Leptocera heteroneura*

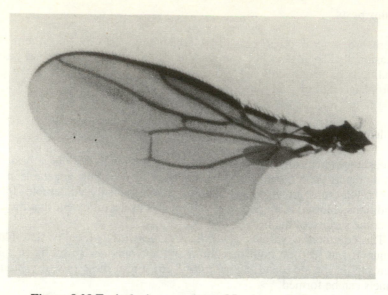

**Figure 8.28** Typical wing venations of *Leptocera heteroneura*

CONTROL

An efficient pasteurization will kill springtails which abound in the surface layers of manure during phase I composting. Cropping-house floors should be kept free of all organic debris and, if possible, crops should be raised off the floor.

## Other flies:

### Sphaeroceridae

Some species of this family, which superficially resemble phorids, occur in manure during phase I composting *(Figure 8.27)*. The most common species to be found in mushroom houses *(Leptocera heteroneura)*, which can be differentiated from mushroom phorids by the red colour of their eyes and their unique wing venation *(Figure 8.28)*, breeds in compost which is unsuitable for mushroom mycelium and where intense bacterial action is taking place. The flies are vectors of mites and pathogenic fungi.

CONTROL

Efficient composting and pasteurization should entirely prevent conditions which are suitable for sphaerocerid increase. However, the normal control methods used for other pests will also prevent their increase.

## Drosophilidae

Drosophilid flies are normally associated with decomposing vegetable matter. However, mushrooms can occasionally be tunnelled by scores of larvae (normally *Drosophila funebris*), and can be reduced to an almost fluid state by the active larvae. Eggs are laid just below the surface of the cap tissues, causing a brown spot which, when it expands, resembles an early stage of bacterial blotch.

CONTROL

Adult flies will be killed by the pesticides used for sciarid and control but, as these insects are opportunistic pests, no specific control measures are justified.

## Scatopsidae

The larvae of these flies, which are usually found in manure or decaying vegetable matter, can breed in mushroom compost where bacterial decomposition is occurring. They are not thought to damage mushroom crops.

CONTROL

Adult flies should be killed by the pesticides used for sciarid and control. However, if conditions which promote putrefying compost are avoided, scatopsids will not be troublesome.

## Diplopoda (millipedes)

Millipedes, which can eat holes in the base of mushroom stipes, are rare in modern tray culture, but may occur on crops grown on the ground.

CONTROL

Apart from keeping the cropping-house floors clear from all organic debris, no other control measures are justified.

## Gasteropoda (slugs)

Slugs, which may cause trouble by eating large cavities in the caps and stalks of mushrooms, are rare and cause problems only in structures which permit their entry.

CONTROL

Metaldehyde baits can be used away from the beds but, in general, no other control measures are justified.

# 9

# Abiotic Disorders

## Introduction

There are many different disorders, and the causes of some are completely unknown. It is helpful for a mushroom producer to be familiar with these problems in order to be able to differentiate between such disorders and the biotic disorders that are caused by specific pests or pathogens. The wide range of symptoms and the infrequent incidence of abiotic disorders is, perhaps, the reason why the cause of many of them is poorly understood. In general, abiotic disorders do not constitute a significant cause of crop loss.

## Distortion

Many forms of distortion can occur; most are probably induced by fluctuating or unsatisfactory environmental factors; they may also be strain-linked. If they occur, the only recourse is to examine the past environmental conditions for some aberration. The following symptoms have been seen.

### HOLLOW STEMS

The mushrooms tend to be of poor quality, but on the bed appear to be normal. When they are cut for market, they are found to have partially hollow stems, often with a circular cavity surrounding a solid core. The cut surface of the stipe then may split and curl backwards, giving the mushroom an unattractive appearance.

## SPLIT STIPES

This is probably an extreme expression of the previous disorder. The stipe begins to split at cutting, or even before (*Figure 9.1a*). Vertical splits may appear and the outer layers of the stem bulge outwards to give a Chinese lantern effect. The vertical splits are sometimes associated with horizontal ones which enable strips of stem to curl upwards, downwards, or both (*Figure 9.1b*). It has been suggested that these symptoms are associated with periodic water stress or fluctuating watering regimes. Chance observations of experimental crops have endorsed these suggestions. Crops being run with very wet casing were allowed to dry and were then brought back rapidly to the high, maintained moisture levels. This process induced symptoms similar to those described.

## SWOLLEN STEMS

Occasionally, mushroom stems can be observed that have large, even swellings so that, instead of the roughly cylindrical shape of a normal stem, there is a concentric bulge. These swellings may be at the base of the stem, in the middle, or at the top. Only in the latter case are the marketable mushrooms affected (*Figure 9.2a, b*).

**Figure 9.1a** Mushrooms viewed from below showing splitting stipes

**Figure 9.1b** Mushroom demonstrating both vertical and horizontal splitting

**Figure 9.2a** Swollen stems viewed from the side

**Figure 9.2b** Mushroom with a swollen stem viewed from below

## HARD GILL

The affected open mushrooms, when viewed from below, have a pale colour and the gills are very shallow, or even non-existent (*Figure 9.3*). If the cap is broken, it often seems unusually thick and it is then apparent that there is little or no gill tissue. Environmental factors are thought to be at least partly responsible for the symptoms, but there is also a tendency for some strains to be affected more than others. Hard gill is known to occur in all strains.

**Figure 9.3** A range of symptoms; from left to right: distortion, split stipe, hard gill

## MISSHAPEN MUSHROOMS

There are a range of other disorders more accurately described as simply 'distortion' (*Figure 9.4*). The sporophores appear to be inconclusively differentiated, with the symptoms appearing to be both strain-linked and most commonly associated with first flushes of autumn crops. The mushrooms, which fail to develop the normal shape, can range from rather knobbly lumps to recognizable mushrooms with rather grotesque misshapen caps. Fusing of individual mushrooms can also occur.

Off-white strains commonly showed the symptoms after their initial introduction to the industry, and similar effects have also been seen on the recently introduced hybrid strains. In the case of the off-white strains, a characteristic symptom, in addition to the more general distortion, is a deep cleft in the surface of the cap. Such distortion can be seen on all strains, which suggests that incorrect environmental conditions (for the strain) are at least part of the cause. Fortunately, the problem tends to be transitory and, usually, does not affect a large proportion of the crop.

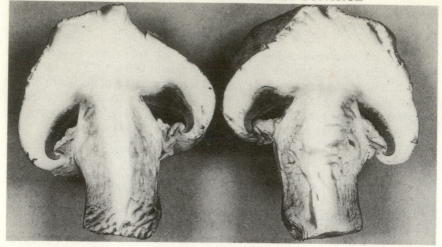

Figure 9.4 Cross-section of misshapen mushroom

'WILD' MUSHROOMS

The almost universal uptake of hybrid strains has changed the incidence of different distortion types, but with hard gill and misshapes being more common. The wide use of these strains has also enabled identification of a new category of distortion. It has been described in several ways, but the most common designation is 'wild' mushrooms. The symptom is a thin, slightly irregular cap, borne on a short and apparently normal stem. This type of distortion may have occurred previously with white (as opposed to white x rough-white hybrids) strains, but would have been less obvious with naturally thin capped, thin stemmed varieties. Amidst thick, domed caps on wide stems, these unnaturally thin capped, casing hugging mushrooms are very obvious. The phenomenon occurs periodically, but not universally, and what evidence there is points to a strain-linked environmental response.

## Waterlogging

There is a range of disorders, the causes of which are uncertain, where waterlogging of the mushroom is a common factor. Symptoms include clear, water-soaked areas, particularly in the stems; the exudation of water from the mushrooms if handled or squeezed; and, in extreme cases, a spontaneous release of large quantities of clear, or coloured, liquid from mature mushrooms, which subsequently collapse. It is possible that, in some circumstances, similar symptoms may be an expression of a viral or bacterial disease. Similar symptoms have also been observed in close association with areas of compost colonized with *Trichoderma*.

## Pinning disorders

There are a range of pinning disorders with more or less identifiable causes. These problems are both more common, and often more devastating, than previously described disorders as they affect both the management of harvesting and, often, the total yield.

### MASS PINNING

This disorder may be recognized when an unusually large number of pin-heads are formed. The result of the phenomenon ranges from the production of a large number of small mushrooms, to the extreme when virtually none of the pin-heads develop and the resultant layer of partially differentiated mushroom tissue appears to preclude the production of mature sporophores. This extreme is similar to a disorder described as stroma.

Mass pinning occurs when the environment is over-conducive to pin-head formation. The so-called normal or prescribed conditions usually ensure the death of many pins, allowing the remainder to develop. The environment that the crop has actually experienced must be examined to identify any aberrations that may have occurred, whether it be in temperature, humidity or watering regimes.

Another, but not clearly proven, technique is to consider taking action specifically designed to reduce pin-head numbers, e.g. heavier watering or higher temperatures.

Mass pinning is now less common, probably partly due to the more reserved pinning characteristics of hybrid strains but, almost certainly, due also to the greater precision in environmental control currently practised. This same precision is more likely to produce subtler problems of under or over pinning, rather than the grosser effect of mass pinning.

### CLUMPING

This problem should be distinguished from uneven flushing, which often has a more apparent and different cause. Clumping normally refers to the occurrence of large mounds of mushrooms, which cause havoc to picking rates and mushroom quality. Some strains are more prone to the disorder than others. There are well-documented instances of periodic low temperatures at pinning having caused the problem, but it is not certain that this is the only cause.

STROMA

When the mycelium grows through the casing and appears on the surface to form a mat, it is called 'stroma'. The term, as a disorder, should more specifically be restricted to the white mycelial capping on the casing surface. This cap is impervious to water, causing waterlogging of the casing surface and gradual drying of the lower layers; where it occurs, cropping is prevented. The cause is probably a combination of high carbon dioxide concentration, high relative humidity and, sometimes, high temperature. Apart from rectifying the environmental conditions, the only recourse is to ruffle (lightly rake) and lightly recase as soon as the problem begins to develop. The disorder is rarely seen now, being more commonly associated with white strains and poor environmental control systems.

PIN DEATH

Pin-heads may die even though the casing is well colonized, giving rise to non-cropping patches. These are easily distinguished from bare patches in which the myceluim has not colonized. The symptom is usually seen in later flushes, and is often caused by waterlogging of the surface due to the presence of an impervious mycelial layer in the casing. This, perversely, is normally attributed to underwatering in the early life of the crop, enabling dense mycelial colonization of the drying casing. Some strains, notably the off-white ones, are less tolerant of this surface overwatering than others.

A similar, but usually more widespread, symptom can be caused by very high temperatures.

In all cases of pin death, pests and pathogens (e.g. mummy disease, virus and sciarid larvae) should be considered because they can cause similar symptoms.

## Carbon dioxide damage

The main symptom is a considerable elongation of the stipes caused by a build-up of carbon dioxide. It is described, in extreme cases, as producing 'drumstick' mushrooms similar to the effects of severe virus disease, but as a disorder it is now rare because of the universal adoption of forced-air ventilation.

## Dirty mushrooms

When mushrooms initiate too deeply in the casing, they often emerge covered in casing material; this is more of a management problem than a disorder.

The nature of the casing mix and picking procedures have large contributions to play, but the essential cause is deep initiation due to either mistiming of initiation or dry casing on the surface. The mycelium does not then grow into the upper layers of the casing, and sporophores are initiated well down in the casing layer. Except on rare occasions, the problem now only manifests itself in poor leaky houses with primitive ventilation systems.

## Rosecomb

This may be described as a specific distortion where pink gill tissue, often with a porous appearance, develops on the surface of the mushroom cap. It may be in warts, or in comb-like structures, or may appear to have spread over the side of the cap from the gills. Mushrooms so affected are grotesque and unsaleable and, as the description indicates, the distortion is pink or rose in colour (*Figure 9.5a, b, Colour Plate 18*).

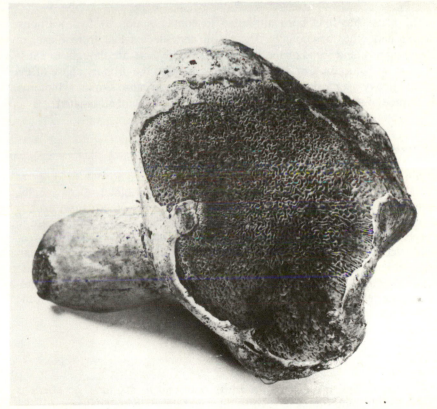

**Figure 9.5a** Rosecomb

**Figure 9.5b** Rosecomb

The cause has long been attributed to contamination by hydrocarbons, phenols and other compounds. Diesel oil and the exhaust from diesel or petrol engines, and some of the ingredients of creosote are thought to cause this type of distortion. Contaminated casing is often the primary cause of the trouble. Heavy overdosing with certain pesticides can cause similar symptoms. The source of the distorting material must be identified and eliminated.

## Browning

Staining of mushroom caps can normally be attributed to one or other of several pathogenic disorders, most commonly bacterial blotch. Occasionally, the nature and incidence of the staining suggests some other cause, usually because the staining is light in colour and diffuse. Should this be the case, it is probable that the cause is chemical. Leakage of phenolic vapour from areas disinfected or the incorrect use of pesticides should be considered. A relatively common cause is a combination of high humidity and the frequent use of over-strong sodium pentachlorophenoxide for tray or bed-board treatments. Enough phenolic material is leached out to drip on to and brown the mushrooms, especially those near the edges of boxes.

## Scaling

This is the natural reaction of the mushroom cap to dry air. The effect is strain-related; some strains (e.g. off-whites and those developed primarily for cave culture) being more prone than others. Air with a low relative humidity

moving over the mushrooms causes scaling: the relative humidities concerned are all above 80% but generally below 95%. Low humidities at very
low air speeds do not cause the disorder, nor do high-speed air movements
with very high relative humidities; the problem arises when the speed is out
of ratio with the humidity.

A further cause is sodium pentachlorophenoxide fumes from trays. Here,
scaling occurs at lower air speeds or higher humidities than would otherwise
be expected. A clue to this cause is the characteristic brown edges to the
scales, which do not normally occur in cases of simple over-drying.

If scaling is a persistent problem, the strains grown, the air distribution
and humidification systems, and the possibility of pesticide damage must all
be considered. The simple expedient of changing to a less sensitive strain can
sometimes eradicate the problem.

### Crypto-mummy disease

A disorder exhibiting symptoms not unlike those caused by mummy disease
has been observed. Its occurrence is quite widespread and its symptoms are
sufficiently similar to warrant its inclusion as a specific disorder. The symptoms which distinguish it from mummy disease are the ability to recover to
produce healthy sporophores and its lack of rapid spread, although it must be
accepted that this particular symptom is difficult to observe with true mummy
disease in tray systems.

The cause of the disorder must be somewhat speculative but studies
strongly indicate that the common factor is chronic over-watering. Increased
levels of intensity of production require greater applications of water, usually
resulting in high yields of good quality mushrooms: however, in certain
circumstances, which involve complex strain and environmental interactions,
it seems likely that chronic excesses of water produce this mummy-like
symptom. Drastic reduction in the application of water, while not bringing
about a complete recovery, often results in the reappearance of healthy
sporophores. Thus, when mummy disease is suspected, not only virus disease
but also crypto-mummy disease must be eliminated as possible causes.

# Glossary of Terms

The following is a glossary of terms as they are used in this book.


**Abiotic disorder** A disorder caused by a physical, chemical or environmental factor (cf. biotic disorder).

**Agar gel column** Agar gel columns are used in analytical processes where it is necessary to separate chemicals with different sized molecules. The process is used in the PAGE test for virus identification.

**Aleuriospores** Cells formed from fungal mycelium which develop thick walls and are able to withstand adverse conditions; aleuriospores are formed terminally.

**Ambient temperature** The natural (outside) temperature at the time.

**Anaerobic** Without oxygen.

**Anastomosis** The growing together of fungal mycelium and the fusion of the cell walls in such a way that cell contents can move from one individual to the other.

**Antenna** A (usually) much jointed, whip-like, mobile, sensory appendage to the head of an insect.

**Antiserum** Proteinaceous materials (antibodies) with properties of chemically recognizing other specific proteins (antigens).

**Apothecia** The sexual spore-producing structures of a group of fungi called the Ascomycotina.

**Ascocarps** Structures in which spores are produced as a result of sexual reproduction in a group of fungi called the Ascomycotina.

**Ascospores** Spores produced in ascocarps.

**Bacilliform** Sausage-shaped.

**Biocontrol, biological control** Usually refers to the control of pests or disease using a biological means rather than chemical.

**Biotic** Disorder caused by organisms and viruses.

**Button** A button mushroom has reached the developmental stage where the cap and stalk are fully differentiated and partly expanded but the tissue covering the gills is not broken, so that they remain enclosed.

**Casing** A mixture of peat and limestone which is placed on top of a compost after colonization by the mushroom mycelium.

**Chlamydospores** Cells formed from fungal mycelium which develop thick walls and are able to withstand adverse conditions (cf. aleuriospores, which are terminally formed chlamydospores).

**Conidia** Asexual spores produced by many fungi.

**Enzyme** A proteinaceous material capable of triggering or catalysing a biochemical process.

**Gills** The tissue in the mushroom sporophore which produces the spores (basidiospores) of the mushroom.

**Hyphae** The individual filaments of the fungal mycelium.

**Initials** Small aggregates of mushroom mycelium which are the first stages of sporophore formation.

**Inoculum** An introduced quantity of a pathogen.

**Knockdown** Term applied to a rapid kill of flying insects.

**Larva** The pre-adult or immature stage of a pest; maggot-like in insects.

**Latency or latent infection** The condition with mushroom virus disease when very low numbers of virus particles are detected but affected crops show no symptoms or yield reduction.

**Latent period** The time between infection and symptom production.

**Lesion** A defined area of decay.

**Macroscopic** Seen by naked eye without any artificial aid to vision.

**Micron (μm)** A millionth of a metre.

**Microscopic** Visible only with the use of a lens or microscope.

**Mould** In this book the term 'mould' is used to describe fungi, other than the cultivated mushroom and its fungal pathogens, which occur in compost or casing at any stage of cropping.

**Mucilage** A sticky material, usually of a carbohydrate nature, which frequently surrounds spores of fungi.

**Mycelium** The vegetative body of a fungus, made up of hyphae.

**Mycophagous** Feeding on mycelium.

**Mycoplasma** Microscopic structures of variable shape capable of inducing diseases in plants and animals, but so far not found in mushrooms.

**Nanometre (nm)** One millionth of a millimetre.

**Over-composting** The condition of the compost where decomposition has proceeded beyond the optimum point for mushroom growth and has resulted in excessive breakdown of the organic material.

**Oviposition** Egg laying.

**Paedogenesis** Asexual reproduction in the larval or pupal state.

**Phytotoxic** Toxic to plants and fungi.

**Pileus** The cap of the mushroom.

**Pin-head** A stage in the development of the mushroom at, or near the end of, differentiation of the cap, but before any enlargement has taken place.

**Polyphage** An organism that feeds on a wide range of other organisms.

**ppm** Parts per million, as a measure of concentration.

**Pupa** Quiescent, non-feeding stage during which a larva changes into an adult insect.

**Relative humidity** The degree of saturation of air at a specified temperature in relation to the total saturation possible at the same temperature.

**Resistant** Normally used to describe a pest or pathogen which is able to withstand what would be lethal pesticides in susceptible or sensitive populations; sometimes used to designate a host which is only partially affected or unaffected by a particular disease, commonly damaging to susceptible strains.

**Rickettsia** A type of bacterium. Those associated with diseases in plants are rickettsia-like and not strictly rickettsias.

**Ruffling** The practice of the disturbance of mushroom mycelial growth in a casing when it has grown about the third of the way up the casing.

**Saprophage** An organism that feeds on non-living organic matter.

**Sclerodermoid mass** An undifferentiated mass of sporophore tissue induced by the presence of a pathogen, in particular *Mycogone*.

**Spawn** Packaged pure culture of a specific strain of *Agaricus bisporus* growing on a medium, usually grain.

**Spawned casing** Spawn-run compost, or specially produced spawn, mixed into the casing before it is applied.

**Spawning** The introduction of spawn into the compost.

**Sporangiophores** The stalks of sporangia.

**Sporangia** An asexual sporing structure of fungi in the *Mastigomycotina*.

**Spore** A general term used to describe various structures produced by fungi which are capable of germinating and reproducing the fungus.

**Sporophore** The spore-producing structure of some fungi, e.g. the mushroom.

**Sticky traps** A card painted with polybutenes, or similar sticky material, on which flies are trapped; used as a method of monitoring pest incidence.

**Stipe** The stalk of the mushroom.

**Strain** A distinct variety or type.

**Supplementation** The addition of a nutritional supplement at spawning or sometimes just before casing.

**Thermal death point** The expression of temperature and time of exposure which will kill an organism (i.e. 60°C or 140°F for 2 hours).

**Under-composting** The condition of the compost where the process of composting has not reached completion and, therefore, the compost is not entirely suitable for mushroom growth.

**Vector** An organism that transmits pests and/or pathogens.

**Verticillate conidiophores** A whorled arrangement of spore-bearing stalks.

**Viroid** Very small ribose nucleic acid particles without a protein coat, capable of causing diseases in plants but not so far found in mushrooms.

**Virus** Small disease causing particles of various shapes which, in the case of mushrooms, have a ribonucleic acid component surrounded by a protein coat.

# Further Reading

ATKINS, F.C. (1974). *Guide to Mushroom Growing.* Faber and Faber, London.

FLEGG, P.B., SPENCER, D.M. and WOOD, D.A. (Ed.) (1985). *The Biology and Technology of the Cultivated Mushroom.* Wiley, Chichester.

FLETCHER, J.T. (1984). *Diseases of Greenhouse Plants.* Longman, London.

GRIENSVEN, L.J.L.D. van (1988). *The Cultivation of Mushrooms.* Darlington Mushroom Laboratories Limited, Rustington, Sussex, England.

HUSSEY, N.W., READ, W.H. and HESLING, J.J. (1969). *The Pests of Protected Cultivation.* Arnold, London.

MACCANNA, C. (1985). *Commercial Mushroom Production.* The Agricultural Institute, Dublin.

VEDDER, P.J.C. (1978). *Modern Mushroom Growing.* Educaboek, Culemberg.

WUEST, P.J. (1982). *Pennsylvania State Handbook for Commercial Mushroom Growers.* The Pennsylvania State University, Pennsylvania.

# Index